Emperors of the Deep

Emperors of the Deep

Sharks—The Ocean's Most Mysterious, Most Misunderstood, and Most Important Guardians

William McKeever

HARPER LUXE

An Imprint of HarperCollinsPublishers

HarperCollins books may be purchased for educational, business, or sales promotional use. For information, please e-mail the Special Markets Department at SPsales@harpercollins.com.

FIRST HARPERLUXE EDITION

Photograph on page ii: Ocean sharks by anas sodki/Shutterstock

ISBN: 978-0-06-291162-9

HarperLuxe™ is a trademark of HarperCollins Publishers.

Library of Congress Cataloging-in-Publication Data is available upon request.

19 20 21 22 23 LSC 10 9 8 7 6 5 4 3 2 1

Contents

Emperors of the Deep

Introduction
Man Bites Shark

I remember my first experience in the ocean. About five years old, I was with my dad, anchored to his broad shoulders. He swam us out past the break, which offered me a panoramic view of the beach, the surf, the distant horizon: a new vantage point that heightened the ocean's tremendous beauty.

My dad liked to swim when the surf was strong, with the waves pounding the shoreline, the kind of surf that the lifeguards warn against taking on. My dad always wanted to test his strength against the waves, to see if the crashing wall of water could knock him off his feet. Most of the time he was left standing. Watching him, I tried to do the same thing, just as any boy might try to emulate his father. One day, when I was ten, the waves were surging, and I summoned the courage to test them. The first wave picked me up and tossed me like a woobie in the wash. I somersaulted and rolled inside the wave, helplessly ransomed in its grasp, until it was ready to spit me out. The sensation of being roiled and tossed and turned and thrown around was totally new. Rather than being scared out of the water forever, I held on to the feeling. I could never predict which way the wave was going to send me, an arm this way, a leg that way, a disorienting feeling I continue to associate with the power of nature. Though a constant source of joy and play for millions of people around the world every day, the ocean was always in charge.

My dad and I would go fishing, and the water wasn't always rough. Out at sea in our skiff in Nantucket Sound, south of Cape Cod, we were in prime territory to land big fish. When my father was ready to cast a line, he'd cut the engine, and we'd drift. I remember listening to the sound of the waves lapping the hull of our boat and the wind blowing through my ears as the boat rocked gently in the water. After a day of fishing, a ghostly swaying in my legs followed me to bed. Sometimes, as I looked out over the sea, I would see a dark spot just below the surface, jutting this way and that. Every now and again, I could clearly make out a shark, its beautiful dorsal fin slicing through the water. Once, my father caught a dogfish shark, a skinny bottom-dweller, about two or three feet long. I remember being struck by how vulnerable the shark looked, thrashing on the boat deck, desperate for oxygen and an escape route back into the water. Though the dogfish is edible, I pleaded with my father to let the shark go. My father gave in, and I was glad to see the fish swim away, rejoining, I hoped, its family and all the other mysterious creatures swimming deep below the surface.

Those fishing experiences with my dad deepened my appreciation of the ocean and the majestic creatures that inhabit its unseeable depths. When summer ended and the school year started, I went to the library and read every book I could find about sharks. I wanted to learn about how sharks operate, how they live their lives. I wanted to know what sharks were really like. But every book I picked up portrayed them as killers. On television and on the big screen, too, sharks were cast as man-eaters, out-of-control predators. A character would go swimming in a movie, and a shark would come along and take his leg off. Conversely, dolphins were never portrayed as anything other than pure gold, like an honor student at school, the paradigm of good behavior.

Slowly but surely, the prejudice against sharks started to seep into my consciousness. I respected sharks as much as I respected the ocean. Both were beautiful and powerful and, I understood even at a young age, in full control—I and all the other swimmers and fishermen were simply visitors to the underwater empire, vacationers, day-trippers, tourists. While my reverence for the ocean included a deep appreciation of its unpredictable power, my admiration for sharks was less sincere, tinged with suspicion and an almost irrational fear. Though I knew I was psyching myself out, every time I went into the ocean I couldn't help but worry that a shark attack was just a few yards away.

As I quickly learned, I was not alone in this fear.

Continuing my research of sharks, I came across what many consider the event that instilled in the American psyche our widespread cultural fear of sharks. It occurred in the summer of 1916. Over a horrifying ten-day period, four people were killed in the water, and one was seriously injured. The events triggered mass hysteria and, as far as I could determine, the first extensive shark hunt in history. Victim One was attacked on July 2 at Beach Haven, New Jersey. Four days later, 45 miles north, a second victim was attacked in Sea Girt, New Jersey. Immediately, hundreds of people took to their boats, armed with nets, guns, and dynamite. Many sharks were killed, each one identified as the man-eater. When the next three attacks took place on July 12—this time in Matawan Creek, 70 miles north of Beach Haven—it became clear that none of the captured sharks was responsible. A taxidermist finally claimed to identify the great white culprit after finding a child's shinbone and what appeared to be a human rib in the shark's stomach. Ever since, however, many have questioned the taxidermist's theory, pointing out that because great whites cannot survive in fresh water, a single great white couldn't have killed the three people in Matawan Creek, a freshwater tidal inlet. (A bull shark is the only shark species that can swim in both salt and fresh water.) While it's likely a great white was responsible for the attack in the saltwater Raritan Bay, the freshwater attacks rule out a deranged, man-eating great white on the loose.

And yet, the single-shark conspiracy prevailed, a questionable-at-best theory that nevertheless served as the basis of a bestselling book, the most famous book ever written about sharks: Peter Benchley's *Jaws*. Since its publication in 1974, Benchley's novel has sold 20 million copies and has likely shaped our perception of sharks, and great whites in particular. When Steven Spielberg released his cinematic adaptation two summers later, the film locked into place the public's intractable fear of sharks. *Jaws* sparked a period of great interest in great whites and established the unsubstantiated belief that sharks as a species are nothing more than bloodthirsty man-eaters, apex predators with no other purpose than to kill. The movie certainly had that effect on me. That opening scene—in which a young woman is attacked and mercilessly thrashed about—made me think twice, as the tagline promised, about ever going in the water again. From the summer of '76 on, whenever I stepped into the ocean, the unsettling specter of sharks scuttled across the deep recesses of my mind, a twitch of fear I tried to ignore while splashing in the surf.

Little did I know that sharks have more to fear than humans. And unlike our fear of them, their fear is justified.

A few years ago, I was spending the weekend in Montauk, New York, a tony community in the Hamptons on the far eastern part of Long Island. Wandering around the marina, I noticed a large crowd gathered around one of the many fishing boats docked there. To my shock, everyone was gawking at a number of dead sharks strewn about the wooden boards, remnants of a recently concluded shark tournament. One shark—a blue shark about 8 feet long—caught my attention. The shark was on full display, its white underbelly exposed, bisected by the marine blue of its body's upper half. Its mouth was gaping open, and a hook protruded from the corner of its mouth. Its internal eyelid had drooped over its eye, which reminded me of a sheet placed over a dead man in a morgue, an appropriate image because I couldn't escape the sensation that I was visiting a murder scene. All these dead sharks had been tossed away like garbage. The sheer number of them on the dock that day rivaled the scene's brutality.

As I slowly retreated from the marina, a knot formed in my stomach. I couldn't help but wonder: If fishermen were killing this many sharks during one weekend on Long Island, how many sharks are dying at similar tournaments around the world? No matter how hard I tried, I couldn't shake the feeling that sharks were in trouble—a feeling I confirmed once I started looking into the numbers. While sharks kill an average of four humans a year, humans kill 100 million sharks each year. That is not a typo. Humans kill *100 million sharks each year.*

While recreational fishing is doing considerable harm to the shark population around the world, commercial fishing is decimating it. Sharks are harvested throughout the world for their meat, skin, fins, liver, and cartilage.[1] Specifically sharks are killed as a by-product of the $40 billion tuna industry, which provides the United States with enough tuna to feed every American 2.6 pounds per year. Seventy percent of the world's

tuna is harvested from the Pacific Ocean, where regulatory bodies are unconcerned with or overwhelmed by
sustainable fishing practices. Over thousands of square
miles of blue, thousands of vessels from China and other
countries fish for tuna and sharks in what Greenpeace
describes as an "industry out of control." More than
five thousand authorized longline vessels are currently
operating in the Pacific, and many more aren't authorized. The preferred method of catching tuna is longline fishing, a remarkable and sadistic practice. Fishing
vessels set a single line from the stern of the boat. The
length of this line is staggering, up to 100 miles long.
Attached to this one line, at regular 10-foot intervals,
are baited hooks. A relatively small vessel can set out
thousands upon thousands of them. Fishermen leave
the line alone overnight, reeling in their catch the next
morning. Anything can get caught on that line, including the sharks that are chasing the tuna. In fact, some
fisheries catch *more* sharks than tuna.

And then there are the horrors of "transshipment."
One fishing vessel transfers its catch to another vessel, which then takes the catch to market. This sounds
like an efficient way of doing business, but when the
transshipment happens at sea, far out of reach of inspection, fishermen can hide what's really going on. Illegal catch is regularly smuggled with legal catch, and

when the entire catch finally reaches shore, there's no way of knowing which vessel it came from or how the fish were caught. At the same time, the crew is often made up of indentured servants, and transshipment keeps these exploited men out at sea indefinitely, often for years at a time. Forced to work for minimal pay, which is rationed out over a fishing season, the crew does what it can to survive, regularly resorting to finning captured sharks, then selling the pectoral and trademark dorsal fins on the black market or to China, where the demand for shark-fin soup is insatiable. Literally millions of tons of sharks are captured every year to keep up with this demand.

These ships are truly weapons of mass destruction on the high seas.

While it is well known that sharks are killed for their fins to make Chinese soup, they are also killed for their squalene, a key moisturizing agent in cosmetic products like lipstick. While squalene can be extracted from plants, sourcing it from sharks is easier and significantly cheaper. Squalene fisheries operate primarily in the southeastern Atlantic and western Pacific oceans, where regulations are lax.

Some people were quicker than others to recognize this peril, most notably the man most responsible for instilling a fear of sharks in the culture at large. Years after the success of *Jaws*, author Peter Benchley was scuba diving off Costa Rica when he had an epiphany. Submerged deep beneath the surface, he spotted the corpses of finned sharks littering the bottom of the sea. This scene, which he called one of the most horrifying sights he had ever witnessed, forever changed his life. Sharks were no longer savage leviathans and man-eating monsters; they were now the mutilated victims in this all-too-real horror story. Benchley renounced his contribution to the "momentary spasm of macho shark hunting" and abruptly changed his views on sharks. Reflecting on *Jaws*, Benchley later wrote that "the shark in an updated version could not be the villain." He added that if he were to do it all over again, the shark "would have to be written as the victim, for, worldwide, sharks are much more the oppressed than the oppressors."[2] Benchley could not escape the carnage that ensued in the wake of his book, and up until his death in 2006, he committed himself to changing the negative perception of sharks and ending their killing.

Today, sharks are under the greatest threat in their entire 450-million-year history. Despite the fact that they play an essential role in maintaining the delicate balance of the marine ecosystem and are key indicators of the overall health of our oceans, sharks are not a priority for conservation. In any healthy ecosystem, a balance of components is in place, evolved over eons. Once the sharks go, environmental repercussions will occur everywhere, from the frigid waters of the Arctic Circle to the coral reefs of the tropical Central Pacific. Taking away apex predators typically results in top-down impacts on the ecosystem. Studies have shown that coral reef ecosystems with high numbers of apex predators tend to have greater biodiversity and higher densities of individual species. A healthy coral system, for instance, requires a similarly healthy shark population, which, in turn, balances the herbivores and the fish that prey on them. Without sharks, the entire ecosystem of the coral reef collapses. Moreover, healthy shark populations may aid in the recovery of damaged coral reefs, whose futures are threatened throughout the globe.

Sharks are being decimated around the world by tuna fishing and finning and even local shark tournaments like the one I encountered in the Hamptons. From a biological viewpoint, the shark population isn't built to withstand the onslaught. With long reproductive and gestation periods, sharks simply can't reproduce fast enough to replace the losses. Kill the sharks, and humankind cripples the seas.

The sudden vulnerability of sharks—a species that has survived five extinction-level events, including the one that killed the dinosaurs—calls for a complete reinterpretation of our relationship with them. That's the goal of this book. I want to shift the perception of sharks from cold-blooded underwater predators to evolutionary marvels that play an integral part in maintaining the health of the world's oceans, because this is the only way I know how to protect the species and save the world's oceans. This is a continuation of my work at Safeguard the Seas, the ocean conservancy I founded in 2019 to use activism, advocacy, and education to protect sharks and other fish threatened by man. Contra *Jaws*, I have figured out the only way to change the way we think about sharks is to cast the species in an entirely new light, in all of their underwater glory, and share with a general audience the emerging science of sharks, which is slowly pulling back the curtain on one of the ocean's fiercest predators. Thanks to new technologies that allow scientists and marine biologists around the world to track sharks like never before, we have made great strides in recent years in understanding the species, including their mysterious underwater behavior, their supernatural migratory patterns, their remarkable social ability, and even their secret sex lives.

To understand these new discoveries, I visited the top oceanographic research institutions in the world, interviewing today's leading scientists and conservationists, from Cape Cod to Cape Town, from the Florida Keys to Australia's Shark Bay, with stops in between. These pioneers broke down for me this emerging science of sharks and shared many of the species' most closely guarded secrets. At the same time, I tagged along with Greenpeace activists in Busan, South Korea, to learn about the organization's most recent "bearing witness" campaign and to tour the *Rainbow Warrior*, the organization's iconic ship. I also met with courageous artists and relentless activists around the world who, like me, are trying to change the public's perception of sharks through creative marketing campaigns and daring exhibitions in Miami and Mossel Bay, South Africa. To get a better sense of how the international fishing industry traffic operates, I interviewed six former slaves-at-sea in Cambodia. Closer to home, I spent a day with the self-professed "last great shark hunter" and went to a mako-only shark tournament in Montauk, the Super Bowl of shark contests, to understand why recreational fishermen continue to hunt sharks for sport, despite their declining numbers. And finally, to get a close-up view of the world's most feared and most misunderstood predator, I stepped freely into their environment, emboldened this time to share the water with the species that has fascinated me since I was a child.

While much has been uncovered about sharks in recent years, there is still much to learn about them—or unlearn in the case of our popular misconceptions about them. Though hardly comprehensive, this book is a culmination of my two-year journey, at once a deep dive into the misunderstood world of sharks and an urgent call to protect them, a celebration of sharks as remarkable apex predators, supersensory navigators, and humankind's greatest ally in nature, and—as scientists begin to learn more about them and their behavior— quite possibly the key to unlocking the mysteries of the ocean. That is, if we don't kill them off first.

Though I focus primarily on the Big Four—great white, mako, hammerhead, and tiger—the ocean is home to approximately five hundred species of sharks, from the epaulette shark, which crawls across the surface of reefs, to the 34-ton whale shark, the ocean's gentle giant, which motors along at its own pace, consuming 50 pounds of plankton a day. The diversity of the species comes across in its behavior and appearance, its form and function. The bioluminescent catshark, for instance, which glows green in the darkness off the continental shelf, is a stunning specimen. Greenland sharks, which measure 13 to 16 feet in length and weigh 3,000 pounds, may live more than four hundred years,[3] making them the longest-living vertebrate in the world by at least one hundred years. They are also the slowest-moving shark, traveling at 1.6 miles per hour. And while the bull shark lacks the natural good looks of the sleek thresher, a ballerina of a shark, the bull's internal machinery, which allows it to survive in fresh and salt water, is undoubtedly a thing of beauty. This evolutionary marvel rivals the great white's deep diving and long-distance traveling abilities, the mako's perfectly proportioned cylindrical body, the hammerhead's trademark T-shaped head, and the tensile strength of a tiger's jaw, which can generate a force of 3 tons per square centimeter, equivalent to the weight of two cars.

Sharks have never been considered beautiful creatures. Yet what all the sharks discussed in this book illustrate is that they are indeed beautiful and majestic, emperors and empresses of the deep, every last one. "If everyone were cast in the same mold," Darwin famously wrote, "there would be no such thing as beauty." Because beauty takes shape when one recognizes the diversity of life and comprehends nature's magnificent design, the experience of it touches the soul. Draped in various shapes and sizes lies a beauty to behold like pearls scattered throughout the world's oceans. We just have to open our eyes to see the sharks as if for the first time.

Chapter 1
Searching for Mary Lee

S hortly after dawn on June 17, 2017, Mary Lee disappeared off the coast of Beach Haven, New Jersey, a sleepy beach community 20 miles northeast of Atlantic City.

Concern quickly spread among Mary Lee's 130,000 Twitter followers, a modest but loyal fan base most likely amassed during her numerous high-profile travels around the world: winters in Palm Beach, summers in the Hamptons, and, in between, the occasional jaunt to Bermuda, where she splashed around in the island's cerulean surf. Wherever Mary Lee popped up, a crowd was sure to gather. Curious beachcombers and paparazzi brazenly snapped photos of Mary Lee with their smartphones and speculated about Mary Lee's real age, her sudden fluctuations in weight, and even the rumored father of her reported pregnancy.

So when Mary Lee suddenly vanished, people naturally feared the worst. Dropping off the map wasn't like her. It certainly wasn't the kind of behavior expected by her growing legion of fans or her handlers. Either she had decided to go offline for a few weeks or—as seemed more likely as June turned to July—Mary Lee wasn't ever coming back.

I first heard about Mary Lee, and the growing frenzy surrounding her, from Greg Skomal, a senior fisheries biologist at the Massachusetts Division of Marine Fisheries and head of the Chatham-based Massachusetts Shark Research Program. Skomal, who's been studying great white sharks in Cape Cod since the Reagan administration, couldn't stop talking about her. I can't say I blamed him. As he made abundantly clear over the course of our conversations, it isn't every day you get to track in real time a 16-foot, 3,456-pound great white, certainly not in the waters of the Atlantic.

"The great white shark is still really an open book," Skomal told me in his office at the Shark Research Program, where he and his team examine the migratory patterns and social behavior of great whites. "When you turn on the television and see white sharks doing incredible things, typically it's filmed in the Pacific and

Indian Oceans. Scientists for decades have been studying this species in those areas. White sharks in the Atlantic have always lagged behind in terms of what we know about its biology, its natural history, its ecology. We didn't have any of those hot spots."

That all changed around 2004, according to Skomal, when the Northeast's previously imperiled seal population started to flourish three decades after the Marine Mammal Protection Act of 1972 prohibited local fishermen from killing them for their furs. Once the seals returned to Cape Cod, so did the white shark, the seals' natural predator. The return of white sharks provided scientists like Skomal with an unprecedented opportunity to study them. For the first time, they had regular access to white sharks in the Atlantic Ocean. "If you can find the species you're trying to study," Skomal said, "you can study it."

And if you can study the species, you can start to figure out its biology, its behavior, and, over time, the size of its population—previously unknown data sets that help scientists get a better understanding of the species' overall health and, as Skomal and others are starting to figure out, how the health of the Atlantic is directly related to the well-being of white sharks in its waters.

Skomal was part of the team that caught Mary Lee

off the coast of Cape Cod in September 2012. Under the careful direction of OCEARCH founder Chris Fischer, the former Emmy Award–winning host of ESPN's *Offshore Adventures* program, Skomal and a group of researchers tagged Mary Lee's dorsal fin—the great white's telltale peak—with a Smart Position Only Tag. SPOT, as the tag is also known, allowed OCEARCH to track Mary Lee's movements and, in a technological twist, broadcast her whereabouts to the general public in real time across Twitter, Facebook, and the organization's other social media platforms.

Because collecting data from free-swimming sharks is next to impossible, the catch-and-release work of OCEARCH is essential. If scientists are ever going to understand these apex predators and their role in maintaining the world's oceans, they need to examine them up close. This requires the fishing skills of a world-class angler and the scientific rigor of today's best marine biologists. Watching the operation unfold is a thing of beauty. In the video OCEARCH released of Mary Lee's capture, Fischer spots the shark while patrolling the waters in a skiff named *Contender*. Once he and his crew catch it, Fischer radios ahead to Skomal, who's waiting a few miles away on the *OCEARCH*, a floating at-sea laboratory. "We are all green here," Fischer tells him, calm and collected amid the growing excitement.

"We are a full go. Big, big, big mature female shark. Definitely a five-plus-meter fish."[1]

Towing the shark, Fischer steers the *Contender* toward the *OCEARCH* and docks the shark into a floating pen, a 75,000-pound capacity hydraulic platform that is slowly raised out of the water, until the pen looks like a wood-paneled pool deck extending off the massive vessel. Then, with the precision of a NASCAR pit crew, Fischer and his team get to work. To lessen the shark's nervousness, one man covers the shark's head with a large towel. He then inserts a large hose into the shark's mouth, providing the captured shark with a continuous flow of fresh seawater so it can breathe in the open air. While that man caters to the shark, others frantically start tagging her, measuring and sexing, drawing blood samples, and performing a quick muscle biopsy to identify reproductive habits and, according to the OCEARCH website, assess organic and inorganic contaminants. One man, who I assume drew the short straw, scraped bacteria and other marine parasites from the shark's teeth. Before releasing the shark safely back into the sea, Fischer christened his great white Mary Lee, after his mother. "I didn't know if I was ever going to have the opportunity to name another shark," he later admitted during a press conference.[2]

The entire operation—catch, tag, and release—took

less than fifteen minutes, but those fifteen minutes ultimately produced more than five years' worth of new research about great whites, the ocean's most mysterious and most misunderstood inhabitant.

"What we've done, for the first time in history, is establish a proven method of capturing the ocean's giants and releasing them alive," said Fischer, an intense, solidly built man in his late forties who seems to burn internally with his organization's mission. "And, in between, giving the leading scientists fifteen minutes of access to use all of the latest technology and to do all of the research projects they've ever dreamed of."[3]

To achieve Fischer's stated goal of providing a brighter future for the white sharks, OCEARCH and their collaborators first need to learn the fundamental pieces of the white sharks' life. If scientists are to preserve the great whites, they need to understand their behavior, biology, reproduction, and range movements. This list is by no means exhaustive. One important place to start is with the population size and the health status of the great whites on the East Coast. Another important area to explore is where the sharks mate, give birth, and grow up as juveniles. To this day, no one has ever witnessed a great white giving birth. If a scientist can determine exactly where white sharks mate and mature, for instance, they can petition the government to protect those maritime areas.

Over thirty-two expeditions to date, OCEARCH has worked with 174 scientists from ninety different organizations, and their ongoing tagging program has catalogued 330 sharks, which are added to OCEARCH's open-source database to aid researchers in their work. In exchange for accompanying OCEARCH on one of its expeditions, Fischer requires from scientists and researchers only a single published academic paper. Since its founding in 2007, OCEARCH has inspired at least twenty-two peer-reviewed academic papers, with many more in the works.

Still, as we enter this golden age of shark research, Skomal was quick to remind me, "There's a lot for us to learn."

While Mary Lee was still trackable online, she revealed many of the ocean's long-held secrets, most notably a great white's underwater behavior, which offered Fischer and Skomal tantalizing clues about possible white shark birthing sites. Fischer told the Associated Press in 2013 that he expected Mary Lee to be more of an offshore fish, taking advantage of the Atlantic's bountiful schools and, potentially, hanging out in the North Atlantic right whale's birthing habitat off the Carolinas. Instead, Mary Lee hugged the East Coast, traveling from Cape Cod to Charleston, then up and down the Southeast, swimming between Jacksonville, Florida, and Wilmington, North Carolina, before heading around Cape Hatteras, toward New York, a few miles offshore from Times Square. While Mary Lee didn't pinpoint potential birthing sites, her path did map out maritime areas worth patrolling further for future catch-and-release programs and, over time, petitioning for protection to safeguard the species. Before she could spill any more secrets, however, Mary Lee went off-line, taking with her all this valuable information. The loss of Mary Lee was a setback, but the best approach as far as Fischer, Skomal, and everyone else was concerned was to keep tagging as many white sharks as they could and, as a result, uncover as many of the species' hidden secrets as possible.

On a subsequent expedition, OCEARCH pinned another white shark—this one named Luci—with a sophisticated acoustic tag. Working in concert with the navy's underwater sonar system, the tag let them track the depth of Luci's dives. Figuring out a shark's migratory pattern was useful, but monitoring a shark as it ventures nearly a mile below the surface of the ocean would help scientists figure out how deep a shark can dive and, once it's down there, what sharks do at those depths. Using SPOT and the acoustic tag, Fischer and his OCEARCH team were able to piece together, meter by meter, one of Luci's most spectacular dives as she cruised along the East Coast of the United States. Luci's dorsal tag sent out a high-frequency ping to a satellite, which established her position at 600 feet below the ocean's surface. While there is plenty of sunlight at the surface, light quickly fades as the depth slowly absorbs it. Within a few seconds of the initial ping, Luci freely exited the sunlit zone, speeding through the enveloping darkness past 660 feet, where only 1 percent of the light penetrates the increasing density.

In 2014, a scuba diver named Ahmed Abdel Gabr, a former member of the Egyptian army, set the Guinness World Record for the deepest scuba dive. Having spent four years training, he successfully dove 1,000 feet under the Red Sea, proving that humans—or at least one with seventeen years' experience as a diving instructor, three canisters of oxygen strapped to his back, and a support team in tow—could survive a deep-sea immersion. Abdel Gabr's descent took twelve full minutes, while Luci blew past this depth in a matter of seconds.

Past 1,000 feet, water pressure begins to reach staggering proportions. One atmosphere at sea level equals 14.6 pounds of pressure per square inch, the same as the weight of the earth's atmosphere. This means that, on land, each square inch of your body is subjected to a force of 14.6 pounds. Water pressure increases about 1 atmosphere every 33 feet of depth, which at 1,000 feet, translates to 30 atmospheres, or roughly the equivalent of one large, 450-pound, adult pig squatting on every square inch of your body. To withstand water pressure at this depth, German engineers reinforced the hulls of U-boats, which allowed them to dive beyond 1,000 feet, but even these engineering marvels couldn't survive a dive beyond 1,200 feet. As scientists continued to record Luci's underwater descent 500 miles away, they were stunned to see her dive past the level at which World War II submarines imploded. The blue light from the computer monitor flickered onto the scientists' stunned faces as they contemplated the shark's remarkable diving capability. The excitement in the room was palpable. Like early explorers venturing across the Atlantic and Pacific Oceans, they were discovering something entirely new, something truly groundbreaking, a high point in shark research and, for many members of the OCEARCH team, the culmination of a career.

A deep dive creates problems for many animals. For humans, high blood nitrogen pressures can exert a narcotic effect, known as "nitrogen narcosis," which during ascent may lead to nitrogen bubble formation, a phenomenon known as "the bends." Great whites avoid this problem at this depth because they don't have lungs or swim bladders.

Well beneath the sunlit zone, which ends at 660 feet, Luci breached the middle of the twilight zone at 1,750 feet. This bewitching area of the ocean (660 to 3,300 feet) is too dark for photosynthesis. Many underwater creatures live in the dim light here during the day but travel up the water column to hunt at night in the sunlit zone. Sharks are highly suited to this area because, like lions, they can see in the dark, thanks to the tapetum lucidum, a layer of tissue behind the retina that reflects light through the retina a second time, increasing the light available to the eye's photoreceptors. Because Luci's eyes can take in light even in near darkness, she would have still been able to attack unsuspecting prey, if she were so inclined. But she wasn't finished diving. Luci easily beat the maximum depth of a nuclear-powered submarine (1,750 feet) and, 950 feet later, eclipsed the height of the Burj Khalifa in Dubai, the world's tallest building. Fischer and his team watched, dumbfounded.

At 3,000 feet, Luci entered the ocean's dark zone. No light can penetrate this depth. No plants can grow there. The only source of food is the "snow" of waste from above. To the surprise—and celebration—of the scientists tracking Luci's descent from afar, Luci bottomed out at 3,700 feet, where she started to swim laterally, surrounded by total darkness except for the occasional bioluminescence of nearby fish, jellyfish, and crustaceans, which flashed in the water like lightning in the pitch-black sky. On a coral reef, the light is dazzling—if only humans could see it, but we are not capable of registering light at that frequency.[4]

Luci is not the only shark star with such diving capabilities. In New Zealand, the National Institute of Water and Atmospheric Research (NIWA) tagged a 16-foot-long shark they named Shack, which set the world's record for the deepest known dive by a great white shark at 3,900 feet. According to NIWA, Shack "regularly . . . deep dives between 3,200 and 3,900 feet while crossing the Pacific Ocean."[5]

Nature has designed great whites like Luci, Shack, and others to routinely dive to great depths around the world. They inhabit one of the most inhospitable places on the planet. The pressure at this level is staggering, roughly the weight of a grand piano on every square inch of a great white's body. At this depth, Luci was the only living thing with solid mass; every other creature was gelatinous and had strange, translucent appendages. Temperatures at 3,700 feet are equally unfriendly at 35° to 39°F, which Luci combated by generating her own body heat. While Fischer and his team studied the mechanisms of Luci's singular descent, they remained mystified about the reason for the dive. One team member theorized that Luci was hunting a giant squid, a mysterious, deep-ocean dweller. Most squid live at the surface and are only 2 feet long, but at 3,000 feet, they grow twenty times bigger. At these depths, squid, worms, and sea crabs grow to monstrous sizes because of a phenomenon known as "abyssal gigantism," a condition scientists link to either the deep-sea environment's higher atmospheric pressure or its colder temperatures. For years, people doubted giant squid really existed. Then some began washing up on beaches around the world. A squid measuring 30 feet in length beached in Galicia, Spain, while other recent discoveries proved that female and male giant squid can grow as long as 43 and 33 feet, respectively. Before OCEARCH confirmed great whites can dive to this depth, the only known predators of giant squid were sperm whales. But as Luci continued to dive, it wasn't too hard to imagine great whites and giant squid waging epic battles at the bottom of the ocean floor.

One of the greatest ironies about white sharks is that they aren't white—or at least not entirely. Only their underbelly is white. This design is shrewd, because in deep-sea water the shark's blue-white countershading camouflages them when pursuing prey.

Great whites are one of the largest carnivorous sharks in the ocean; however, they are only sixth in overall size compared to other sharks, a fact that belies the notion of the great white as a killing leviathan. The largest sharks, which are harmless to humans, belong to the filter feeder category—the whale, the basking, and the megamouth sharks, all of whom are larger than the great white. Like great whites, the Pacific sleeper and Greenland sharks are carnivorous but larger, according to Greg Skomal. Unlike many species, where males are bigger than females, female white sharks like Mary Lee are larger than their male counterparts, which on average measure between 11 and 13 feet and weigh between 1,500 and 1,700 pounds. Mature females grow to 15 or 16 feet and can weigh up to 2,500 pounds. While males can easily reach 17 feet, it is not unusual for a female shark to grow to 20 feet in length and weigh 4,300 pounds, equal to the length and weight of an adult giraffe, as difficult as that is to imagine. A white shark caught in Cuba in 1945 measured 21 feet in length and weighed a staggering 7,300 pounds, or 3.5 tons, a weight equal to six adult grizzly bears. The reason female white sharks are larger than males is simple: as Skomal told me, females need considerable strength to carry their pups during the white shark's eleven-month gestation period, all while continuing to hunt. It takes longer for a white shark to develop in utero than it does a human, mainly because once they are born, shark pups are on their own. Unlike dolphins and orca whales, which protect their babies, white sharks leave their pups to fend for themselves. The great white pups eat what they can catch, which in their infancy is fish. As white sharks mature, they start hunting seals and other larger mammals.

"[Pregnant white sharks] are older animals," Skomal explained. "They're in their twenties and thirties when they reproduce. And they're not capable of reproducing until they hit those sizes and those ages. When an angler removes a young, or small, great white shark from the ocean, it has a significant impact on the population because that shark probably hasn't lived long enough to breed." The population replacement rate for the white shark is extremely low, which makes the species vulnerable to exploitation. "They're maturing at a late age and only giving birth to a handful of young, most likely, every two to three years. We have to be particularly conscious of this when it comes to sustainability, conservation, and management of the species."

While there is still no reliable data about the world's total great white population, scientists all agree that the total number is dropping, largely because of overfishing, hunting, and other environmental factors. White sharks are currently listed as vulnerable, a tick above endangered, on the International Union for Conservation of Nature's Red List of Threatened Species.

The white shark's vulnerability belies the popular misconception of the species as bloodthirsty man-eaters. "When people think of white sharks," Skomal told me, "they think all kinds of things. Most of them are fairly negative, which came out of *Jaws*. Hollywood has done a very good job of scaring the hell out of people."

Jaws is a primary driver of humans' nearly five-decade-long counterattack against sharks, an assault that threatens to endanger the species and, in the process, upset the delicate balance of the marine ecosystem.

A fear of sharks has led people to seek the thrill of catching them. But what I have found out about the great white is extraordinary. In sharks and in life, fear is often the absence of knowledge. "The more people know about these animals," Skomal said, "the more likely they are to revere them as opposed to fear them. The more we're learning about sharks, the more we're learning that they're an integral part of the marine ecosystem." Fischer and Skomal and an entire generation of marine biologists and conservationists have dedicated their careers to trying to change the public's perception of sharks, specifically great whites, as underwater monsters.

Like most teenagers of his generation, Skomal discovered marine life from TV shows like *The Undersea World of Jacques Cousteau* and *National Geographic*, which brought color images of sharks and other underwater wonders to living rooms around the world for the very first time. But what really impacted Skomal were his family vacations to the Caribbean, where he fell in love with the ocean, mesmerized at an impressionable age by coral reefs and the variegated fish species he saw scuttling about their natural environment.

"When I was, like, twelve, thirteen years old, I wanted to study sharks, but I figured by the time I got old enough to do it, it would all be done," he said. "How naive was I?"

Later, after resolving to learn everything he could about the ocean, Skomal enrolled at the University of Rhode Island, where he earned his bachelor's and master's degrees. He later returned to school to earn his PhD at night at Boston University. While searching for a full-time researching job, he volunteered at a federal laboratory. Surrounded by field scientists with years of experience, Skomal conducted scientific investigations, developing in the process an unshakable passion for great whites. In 1987, he landed a full-time job as a senior fisheries biologist at the Massachusetts Division of Marine Fisheries, where he quickly realized that most of the scientific knowledge about great whites emerged from hot spots in the Pacific and Indian Oceans. Fortunately, his career coincided with the return of great whites to the northeastern Atlantic Ocean, once the seal population returned. "I was at the right place at the right time," Skomal told me. The entire Atlantic Ocean was suddenly his to explore and research, uncontested. "Many white sharks come up to Cape Cod in the summertime, and simply move in the wintertime down to the coasts of Florida and Georgia and South Carolina. But then we have a component of the population that wanders the Atlantic, and those are most intriguing to me. You know, not just the coastal migratory pattern, but the ones moving out into the central part of the Atlantic, where they're diving to great depths."

Like Fischer, Skomal has a nose for finding great whites. To date, he and his team have tagged and tracked more than 135 great whites and have identified approximately 300. In a 17-foot skiff, he regularly patrols the waters off Cape Cod's Monomoy Island, an 8-mile-long run of sand extending southwest from Chatham. The island is a popular congregating spot for seals. Often, Skomal films these encounters. From his boat, he looks out for the great white's unmistakable shadow underwater: a dark mass moving through the green water. When he sees a shark, which in some cases is almost equal in size to the boat, he approaches the prow and plunges a tag into the base of the passing shark's dorsal fin. No worse for wear, the shark swims on, unbothered.

As Skomal described it, hunting for sharks sounds routine and uneventful, like swimming laps in a pool. But it isn't always this easy or stress-free. Once, trying to get a close-up of the shark's face to help identify it, he attached his GoPro on a pole and plunged it into the water. He had done this scores of times but on this occasion, an 11-foot female went right for the camera. "It kept coming and then opened its mouth and bit it," he said, calling the shark's action exploratory, rather than predatory. If it had been the latter, he reminded me, the shark would have destroyed the camera and, in all probability, pulled Skomal into the water during their brief tug-of-war.

Skomal added that this shark was not one of the 135 sharks in the area that were already tagged by his team for research. He and his team will now review the video to look at her markings and determine if she's brand-new to the area or if she is one of the 300 sharks his team has tracked previously. These experiences and the research information that come out of this tagging program have helped to answer a number of questions about great whites, including their life expectancy, which has confounded scientists for years.

Previous studies concluded that great whites live into their twenties and thirties. However, as scientists continue to collect more information, such estimates are proving problematic. Schoolchildren know that as trees grow, they lay down rings on an annual basis. Each ring represents a year. Sharks similarly lay down band pairs of rings on their vertebrae on an annual basis. While this trait was known in small to medium-large white sharks in the northwestern Atlantic, what was not known was that, after maturity, the largest sharks may experience a change in the rate of vertebral material deposition. Another difficulty scientists encountered while trying to determine a shark's age was that some bands become too thin to read accurately.

The best scientific method to determine the life expectancy of great whites is radiocarbon dating. This well-known method uses the properties of radiocarbon (carbon-14), a radioactive isotope of carbon, to determine the age of an object. But where could scientists find the white sharks for the test? It just so happened that a lab in Narragansett, Rhode Island, contained the largest collection of vertebrae samples from white sharks caught in the northwestern Atlantic Ocean from 1967 to 2010. Using this material and the National Ocean Sciences Accelerator Mass Spectrometry facility at Woods Hole Oceanographic Institution, scientists were able to determine that great white sharks can live to over seventy years,[6] which means that great whites are alive today that, as pups, heard the sound of US depth charges attacking Nazi submarines in World War II. Based on the data they collected from Mary Lee, Skomal estimated that she is in her early thirties, a woman in her prime. Because she likely has another forty years in her, if she were still online, she would have helped scientists identify her preferred breeding ground and pup nursery, which could have provided Skomal and other scientists with invaluable insights to assist with management of the species.

Tagged great whites appear to give birth in certain areas, although this process is not fully understood. Further complicating the issue is the fact that white sharks are found all over the world, which multiplies potential nursery sites exponentially. Based on tagging programs and circumstantial evidence, however, scientists are beginning to zero in on a few nurseries. Some are believed to be located off Taiwan and Japan. Another possible location is the Sea of Cortez off Mexico, because several tagged females went there from April through August, though scientists were unable to prove this hypothesis.[7] Montauk, Long Island, is likely home to a small nursery, 100 miles east of Manhattan. Some organizations have tagged numerous juvenile great white sharks in this area, indicating that this territory is probably a birthing location. Given the proximity to the Long Island Sound, this site offers great white pups plenty of baitfish. When Mary Lee was spotted near Montauk, rumors abounded that she was about to give birth.

The behavior of white sharks, however, is far from predictable. According to Skomal, there is no such thing as a typical day in the life of a great white. "When I tried to come up with the average day in the life of a white shark, I found that it's really difficult to do," he explained. "Now, we've tagged scores of white sharks in the last six years . . . a fairly respectable sample size for that species. It's an elusive shark; we don't believe its population size is very big. So, our database should give me a nice snapshot of what they're doing. And what we're finding out is they're doing whatever they want. Some white sharks will hang around Cape Cod for the whole summer, and they get into a routine of just basically moving up and down the coastline over the period of three or four months. Other sharks may stop by Cape Cod before moving on up into the Gulf of Maine; other animals might only be there for just a couple of hours before heading off deep into the Atlantic Ocean. Every shark seems to be very different. And I'm not getting any real patterns that tell me what the average day in the life of a white shark is really like."

I asked Skomal how close sharks get to shore. "We've tagged white sharks, literally, within feet of the shoreline . . . almost touching the sand of the beach itself. They are hunting in that very shallow water for seals. So there's no doubt in my mind that they're moving within close proximity of humans, quite possibly routinely, and they have been doing that for hundreds of years."

So much for the great white's reputation as an insatiable underwater assailant hell-bent on killing unsuspecting beachgoers. In fact, a great white's proximity to land makes it more vulnerable to humans. While great whites are classified by the US government as a "prohibited species," commercial and recreational fishermen alike can still catch white sharks as long as they don't keep them. Usually, when a recreational fisherman encounters a white shark, the shark is feasting on a dead whale. Most recreational fishermen are content to film and photograph a great white in action, because they are a difficult and dangerous species to capture.

However, some recreational fishermen do target them. White sharks are hardy animals, but if one is hooked deeply or in its gill, or banged against the side of the boat, the damage can prove lethal. Longline commercial fisheries, on the other hand, inadvertently capture white sharks as bycatch, which for the time being is simply the cost of doing business in the open seas. Because they are prohibited from keeping or selling great whites, they let them go—usually after the sharks are already dead. Skomal described this act as "cryptic mortality." "The species is unquestionably vulnerable to directed exploitation," he said. "Unfortunately, it's unclear how great whites are faring." And there is always the situation where fishermen can get away with murder.

When a great white shark washed up on a beach in Aptos, California, the question became, how did this shark die? The nine-foot-long shark appeared healthy. As the Department of Fish and Wildlife inspected the fish, they noticed three bullet holes from a .22 caliber rifle. The case was solved only through an anonymous tipster who revealed that the shooter was a commercial fisherman, Vinh Pham. Upon questioning, he said that the shark "was disturbing his fishing activity." The punishment for the crime of murdering a great white in cold blood—$5,000 fine and no jail time. He did not even lose his fishing license.[8] As long as our society values one of the world's great apex predators as worth nothing more than a small fine, the killing of great whites will continue.

Aerial surveys suggest that great whites are rebounding off the northeast coast of the United States, and Skomal's work conducting surveys in these waters since 2009 bears this out. In his first year, he spotted only five sharks in the Cape Cod area. Seven years later, in 2016, he spotted approximately 150. Still, the exact population of great whites in the United States remains unknown. Similarly, the International Union for Conservation of Nature (IUCN) can't accurately estimate the total population of great whites around the world, even though it can tabulate the populations of other vulnerable and endangered species, including snow leopards (5,000), tigers (3,000), and black rhinos (4,800). The absence of hard numbers is troubling, because without them, conservationists are unable to come up with a plan to help protect the world's disappearing white shark population. And that population becomes increasingly vulnerable as individual sharks traverse the oceanic hemispheres; great whites like Mary Lee and Luci aren't only record-setting divers, they're also marathon travelers. A look into the past can explain how they became such great swimmers.

The age of the fish began about 530 million years ago, during the Cambrian explosion. Nature kept coming up with new designs, and 450 million years ago during the Silurian period, nature developed the relative to our modern sharks. For the next 150 million years, nature tinkered with, developed, and improved the sharks; evolution adjusted the jaws, molded and rounded the head, and experimented with new shark species. For example, the *Helicoprion* shark grew a table saw–like set of teeth on its lower jaw in the Permian period, 280 million years ago, though it became clear that this variation on the species didn't work. During the Carboniferous period, 300 million years ago, sharks dominated the oceans and split into subspecies like skates and rays. By the Jurassic period, 200 million years ago, the predecessors of today's sharks appeared. New species kept appearing through the ages, like *Hybodus*, which had horns but then went extinct. By 60 million years ago, nature had developed the sharks we recognize today. One of nature's most enduring creatures, the shark's design was extraordinary, allowing it to survive and rule the seas for literally millions of years as one of the world's top apex predators.

The previous blueprint for fish required bones, along with supporting vertebrae, scales for protection from the water, and swim bladders that gave fish their remarkable buoyancy. Because all fish had to do to escape danger was to use its vertebrae to flick its tail for a quick getaway, fish brains were small. Over time, however, nature threw away this blueprint and started all over with the shark. The bones were discarded in favor of cartilage, which offered the shark structure and support, and a new material called dermal denticles replaced scales. Denticles turbocharged the shark's speed in the ocean since they reduced resistance. Over time, the shark's brain grew larger, which allowed for a higher intelligence in sharks. Escaping evolved into hunting. Numerous popular articles have described the brain of a white shark as being the size of a walnut, a misleading and inaccurate comparison. The brain of an adult white shark is shaped like a Y, and from the scent-detecting bulbs to the brain stem, a shark's brain can measure up to approximately 2 feet in length. In comparison, the brain of a human comprises two wizened hemispheres, roughly the size of a head of lettuce. Of course, because large animals tend to have larger brains, a more meaningful comparison is between brain weight versus body weight. The brain of a 1,000-pound great white shark can weigh 35 grams, or about 0.008 percent of its total body weight. In comparison, the human brain weighs 1,400 grams, or 1.9 percent of our total body weight. Relative to the body weight of birds and marsupials, however, the great white's brain is large.[9]

An astonishing structure, the brain of a great white shark is composed of millions of neurons, or nerve cells, which contain supporting structures. The brain coordinates the shark's many movements, from clenching and opening jaws that can either rip prey apart or, if the situation calls for it, delicately grasp an object, to lashing its tail to scare off a competitor. The shark's brain is arranged in a linear fashion. Specialized regions line up like a jeweled necklace, from the brain stem to the posterior cranial nerves, which are responsible for conveying information from the shark's inner ear, lateral line, and electrosensory systems. Moving toward the top of the brain, next is the cerebellum, where sensory inputs come together to help generate movement. A white shark's cerebellum is well developed, which can explain the shark's speed and reflexes. In the shark's midbrain are the optic lobes, which process what the shark sees. A special vessel arrangement near their eyes warms them and the brain for faster processing. Another advantage of this capability is that it helps the shark travel through waters where the temperature changes very quickly.

After the midbrain is the cerebrum, where the shark thinks. In this area of the brain, home-ranging and social behavior occur. Great whites use this part of the brain to identify and track prey, process environmental markers for food sources, and recognize potential mates, to name just a few items. The cerebrum is also where the shark's brain splits into the two cerebral hemispheres, a unique feature among vertebrates. At the top of this Y are the olfactory tracks that the shark uses to smell.[10] Because some 70 percent of the shark's brain is dedicated to this sense, the shark is perpetually enshrouded in a world of scents. The reality is that great whites are intelligent and are endowed with a brain superior to that of the other fish.[11] For instance, salmon have a fraction of the cerebral endowment of a great white.

Denticles, which replaced the scales of fish, became the new skin for the shark. Denticles are essentially modified teeth with an inner core made up of tissue and blood covered by a hard outer layer of calcium carbonate. Each one has its own unique shape, but the basic structure is similar. Think of the design like a bicycle helmet with a round front and three main ridges flowing from front to back. Each ridge tapers into points at the tail end. The denticles are crammed together like overlapping shingles on a roof, covering the shark. If you rub your finger over a shark from head to tail, the denticles feel smooth, but run your finger in the opposite direction, and the skin is rough.

Inspired by the shark's denticles, engineers at Harvard's School of Engineering and Applied Sciences have been studying and testing ways to improve the aerodynamic performance of airfoils, or wings. The engineers took a smooth airfoil and arranged 3-D printed shark denticle devices on its upper surface and investigated the effect on aerodynamic performance. Using a complex software program, engineers performed tests in water tanks and made computational analyses of fluid dynamics. They discovered that the airfoil with the attached shark denticles resulted in the formation of vortices behind the attachment. A short separation bubble appeared in its wake. The denticle is essentially a vortex generator, and these vortices are responsible for up to a 10 percent reduction in drag.[12] The Harvard engineers also discovered that the denticles enhanced lift and even helped to maintain lift at higher angles of attack. Therefore, the shark's denticles simultaneously enhance lift and reduce drag, resulting in large lift-to-drag ratios.[13]

As this explains the shark's lightning speed in the water, engineers are looking into copying the shark's design and applying it to any wing on a plane, helicopter, or other aircraft. Engineers can also use the design on wind turbines to enhance their performance. In coming years, different wing shapes may appear with improved performance, and if so, society will need to thank the sharks for providing the design's inspiration.

Besides the brain and denticles, nature made other improvements to the shark, and a significant one was the liver, which holds an oil reserve that helps sharks stay afloat and traverse long distances. Through evolution, fish came to rely on a swim bladder for buoyancy, which prevents them from wasting too much energy. A fish's swim bladder is usually two gas-filled sacs located in its dorsal portion. However, without a swim bladder, sharks still required something to ensure buoyancy. Working double duty, an oil reserve in the shark's liver solved the flotation problem and provided sharks with the energy to propel themselves through the water for long distances.

But other developments beyond the shark's internal oil reserve made long-distance travel possible. The shark has pectoral fins that stick out from its side like wings on a plane. The shark uses its tail to move forward but uses its pectoral fins to pitch up or down. Although the oil in its liver allows for buoyance, most sharks have a negative buoyancy, which means that because their bodies are denser than the fluid they replace, they have a natural tendency to sink if they're not moving forward. The great white, however, turned this seeming disadvantage into an advantage. When great whites begin their trip with a slight downward orientation of the pectoral fins and the tail for propulsion, the negative buoyancy allows the shark to simply glide downward with minimal effort. After reaching a certain depth, the shark makes an upward adjustment with its pectoral fins and, as its tail powers the shark along, it can once again ascend. This type of swimming is known as "drift diving," which makes the shark very effective at traveling long distances, since drift diving requires far less energy—sometimes 50 percent less than the energy required for swimming forward at a specific depth.[14]

Mary Lee, Luci, and their fellow great whites make staggeringly long voyages around the world this way. Great whites have been tracked going from Mexico to the Hawaiian Islands and back again. As any car owner knows, a trip of this length would require a lot of stops at the gas station. In fact, a typical car would have to stop and fill up the gas tank thirty-seven times for a comparable journey. Where can the sharks pull over and refuel? As it turns out, the sharks don't hunt on these migrations, so they rely on internal stores of energy. Unlike whales and terrestrial animals that can draw from energy stored in blubber or fat during long-distance migrations, sharks don't carry blubber; they bulge with muscles, like Olympic swimmers. To travel long distances, sharks rely on their body oil, which is held in the liver like a giant storage tank. A shark's liver, which sits in the abdominal cavity, is huge, extending roughly from the shark's esophagus to its pelvic fin. Oil in the liver accounts for a quarter of the weight of a great white, which means a 2,000-pound great white is carrying 500 pounds of oil, or roughly 60 gallons of oil, more than twice the fuel-tank capacity in a Cadillac Escalade.

Before NIWA started tagging sharks in 2012, they thought that great white sharks lived in cold water only. But after five years of diligent work, they now know that great whites in New Zealand migrate to tropical waters in winter, abandoning the area between April and September for warmer temperatures in the north. In migrating these distances, Shack and other great whites confirm they are remarkable travelers and superb at long-distance migrations that can match the travels of whale species like the grays and humpbacks. The maximum distance one New Zealand shark migrated in winter was 2,000 miles. The tag data reveal that great whites routinely travel 100 miles a day, whereas humans walk an average of 2.5 miles a day.

Now that science can track where the sharks travel, the information can be of great importance in helping protect and manage shark populations. Perhaps the best way to understand this is through the life-and-death experiences of whales in the Stellwagen Bank National Marine Sanctuary, 25 miles east of Boston, between Cape Ann and Cape Cod in Massachusetts Bay. This 842-square-mile federally protected marine sanctuary is a safe haven for whales and other marine species. When scientists were tracking humpback and right whales in the sanctuary, however, they discovered that the whales' path crossed against ships entering and leaving Boston Harbor. As a result, whales were being killed by ship strikes. Authorities changed the shipping lanes, increasing shipping costs but dramatically reducing whale fatalities. The same analogy applies here to the sharks. By knowing the migrations of tagged sharks, we can use emerging information about mating and breeding areas to protect the species.

A similar rush to understand great whites and their migrations is underway in the Pacific, where great whites feast on a rich diet of blubbery elephant seals and sea lions along California's central coast. For reasons that remain unclear, however, great whites make a strange migration in the spring. Like sailors following a siren call, the sharks leave behind the coast's rich cornucopia of blubber for a patch of territory more than 1,000 miles away, halfway between the Baja Peninsula and Hawaii.[15] Nicknamed the White Shark Café, this area, which is approximately the size of the state of Colorado, hosts what some people have started to refer to as Burning Man for sharks. One plausible explanation for the sharks' mysterious journey is sex. In some species, females visit an area to find a mate in what is called "lekking." (A "lek" is an aggregation of males.) In the Café, where the Pacific's chlorophyll-low waters offer great clarity, female white sharks check out male sharks, specifically their fins and muscle tone. To show off, male sharks execute rapid oscillating diving patterns known as "bounce dives," which require great strength and stamina. The shark will dive at night 500 feet straight down before returning back to the surface, creating a birdlike V-shaped pattern in the water. During the day, sharks increase these dives to 1,500 feet below the surface. One industrious male completed ninety-six dives in a single twenty-four-hour period, and the males keep up this behavior for three months inside the Café. The females watch this behavior and select the most prepossessing male. At the same time, the males might be moving at various depths to find the female pheromones, which they can track to their source and make a display of beautiful dives to woo the female.

Like most other reasons for doing something, if it isn't for sex, it's probably for food. It's entirely possible female white sharks headed there as culinary tourists, innocently engaging in foraging behavior, until the males showed up and turned the Café into a pickup bar. Still, scientists point out that the females do not make the same bounce dives as the males, and though humans have explored only 5 percent of the world's oceans, the proponents of the sex theory are confident that the Café isn't home to a unique food resource that would draw females back and forth from California, an exhausting round trip totaling 5,000 miles, approximately 700 miles farther than the wildebeest's annual grass-munching trek across eastern Africa. At this stage in the research, no definitive conclusion has been reached.[16] Foraging, mating, and—like their Burning Man counterparts—communing as one species all remain under consideration. Perhaps it's all three at once.

To better understand the sharks in this area—and to protect great whites on the high seas—scientists have descended on this site with an armada of ships and tools. Reflecting the importance of this area, a 2016 UNESCO/IUCN report identified the White Shark Café as a potential World Heritage Site.[17] If this site is approved, great whites will have an area protected from fishing vessels, which will give the species a better chance for survival. Of course, as we'll learn later, fishing fleets from around the world want to exploit this area of the Pacific. Given the increased vulner-ability of the great white, it would be a double tragedy if fishing activities in protected areas interfered with the mating of the species. Like the work in the Atlantic with Mary Lee, the more knowledge society has about great whites, the greater the likelihood that the pro-posed fishing regulations will be effective. The race is on between the industrial fishing fleets of the world and the scientists to unravel the mysteries in order to implement the optimum regulatory fishing decisions.

In the meantime, scientists like Greg Skomal con-tinue to tag and track great whites to gather as much information as they can, hoping a currently unknown shark, just waiting to be discovered, can offer unimag-ined insights—just as Mary Lee did before she went offline.

Mary Lee's disappearance remains a mystery, and it always will, though Chris Fischer doesn't believe a commercial fishing vessel got her. Nor does he worry about a recreational angler cosplaying Quint from *Jaws*. Neither scenario is likely, he said. "Mary Lee's the queen of the ocean. She's a mature white shark that absolutely dominates wherever she goes." Fischer and Skomal both believe that, after five years, the battery in Mary Lee's tag simply ran out of juice. Like them,

I often imagine Mary Lee is still out there. Perhaps she found a mate and delivered another litter of pups, somewhere off Montauk—close to the likes of Julianne Moore, Robert De Niro, and other A-list celebrities in the Hamptons, who are likely unaware of the royalty swimming in the waters nearby: the great white whose secrets revealed to scientists how to start safeguarding the ocean and its underwater denizens for future generations.

Chapter 2
Makos, the F-35 of Sharks

The Woods Hole Oceanographic Institution (WHOI) is the largest independent oceanographic research institution in the United States, with more than a thousand staff members. Established in 1930 in Woods Hole, Massachusetts, WHOI operates ships around the world, where scientists and engineers work to understand the ocean and its relationship to the rest of the planet. I went there to meet Jelle Atema, an expert in shark sense.

Born in the Netherlands, Atema whittled flutes out of wood as a boy and observed animals in the nearby woods, developing at an early age a love of music and nature, neither of which he's been able to shake. Later, he studied biology at Utrecht University, one of the oldest universities in the Netherlands, before earning his PhD at the University of Michigan, Ann Arbor, in 1970. Four years later, he joined the faculty at Boston University's Department of Biology and Marine Program, where he matured into a tenured professor, internationally renowned biologist, and, in his spare time, a world-class flutist, performing at venues throughout the United States, Europe, and Asia. It's no wonder, then, that Atema now goes by the nickname "The Original," a sobriquet he picked up in 2017, when attendees at an annual ocean conference noted that Atema was the only researcher present who had attended the inaugural conference forty years earlier.

Atema has published 175 papers at the WHOI—all of which have focused on how aquatic animals employ their underwater senses. Through the years, sharks have proved to be a particular interest of his, mostly because the species is constantly in search of prey, which makes their individual senses an invaluable source of information. Like most animals, sharks rely on their acute senses of scent and sight, taste and hearing, to move about. But unlike other animals, sharks are equipped with something called "flow detection," a keen ability to track trace odors left in the wake of a prey. Though it's long been rumored that a shark can detect a single drop of blood a mile away, Atema told me this overstates the powers of a shark's olfactory system—not to mention the primary laws of physics. For the odor of a single drop of blood to reach a shark, it would have to remain intact. Anyone who's ever cut themselves shaving knows that blood quickly dissolves in water. Still, Atema told me, "if a lot of blood gets into the water, and a moderately diluted drop of it reaches the shark a mile away, the shark can probably locate the source" by following the flow patterns across 5,280 feet underwater. Odor alone, though, isn't enough for a predator to locate its prey, because odors can't indicate the direction of their origin. However, as animals move through the water, they leave behind an odor trail, inadvertently scenting the water in much the same way humans and other animals coat the air with their respective odors when they move across land. Called "odor plumes," these flavor-scented eddies are complex three-dimensional structures that attach to water particles. Picture the oily residue in the wake of a moving boat. Many sea animals use these swirling eddies of water to locate prey, potential mates, and, when they're ready to call it a day, the location of their homes. Because odor reaches each nostril at different times, sharks and other aquatic animals can determine the location of the plume and thus which direction to swim toward.

In addition to their noses, sharks use their skin

to find prey. A shark's lateral line, a racing stripe of an organ that runs the length of a shark's body and across its head, has evolved with breathtaking accuracy. One can make out the thin lines in profile photos of sharks. With exquisite sensitivity, the lines can detect movement, vibration, and pressure gradients in the surrounding seawater. Filled with fluid, the canal-like lateral lines are sheathed with epithelial cells with short hairs called "cilia" that sway within the fluid when they come in contact with water motion. Sharks can detect vibrations as low as 1 hertz. The number of hertz equals the number of cycles per second. The human hearing organ operates in a higher bandwidth at roughly 20 to 20,000 hertz and peaks in the 300 hertz range while sharks have greater sensitivity and can operate at a lower bandwidth. Another way of stating this capability is that sharks can detect motion as minuscule as the width of an atom. Lateral lines convert the movement of the cilia into electrical impulses. The shark's brain processes this information to give directional information.

To test the accuracy of lateral lines, a German experimenter manufactured a trail in a swimming pool using a toy boat and set a trained seal after it. (While seals lack a lateral line, they do have whiskers, and lateral lines and whiskers work in a similar manner.) The manufactured trail featured a 90-degree angle—a hard right-hand turn. In a burst of creativity, the experimenter blindfolded the seal, hypothesizing that, even hunting blind, the seal would stay locked on to the turbulence of the trail through the turn. It did.[1] Like the inspired experimenter, Atema has spent years conducting original research about the uncanny underwater sensing abilities of aquatic animals. He started with sharks and then moved on to lobsters and other aquatic animals. In his shark research over the past fifteen years, he has focused his work at WHOI and Boston University, trying to figure out exactly how sharks employ lateral lines to sense and locate prey, eventually proving that sharks rely on the deadly one-two punch of their lateral lines and olfactory senses to hunt. He published his findings in a seminal 2014 paper, which included a schematic of how shark senses operate, separately and in tandem, during a hunt.[2]

Atema was able to piece together why sharks make mistakes and on rare occasions accidentally bite humans. A shark will make visual contact with its prey before striking. Underwater, visibility is limited and varies considerably based on conditions. On sunny days, for instance, when the water is clear, a shark can see the length of a football field. Conversely, on cloudy days, when the water is churned and murky, a shark might only be able to see a few feet in front of its nose. Under these conditions, a shark may not be able to know precisely what it is attacking, which explains why sharks sometimes mistake surfers for seals.

Sharks have the ability to sense small changes in electrical fields in the water. This sense, called "electroreception," is vital to understanding sharks' remarkable hunting skills. Electroreception occurs in organs called the "ampullae of Lorenzini," a series of dark pores leading to canals (ampullae) that are scattered under the skin across a shark's head. They allow the shark to sense electrical pulses of fish. Named after Stefano Lorenzini, an Italian physician and noted ichthyologist who examined shark cadavers in seventeenth-century Florence, these pores reminded Lorenzini of canals (*ampullae* in Latin), hence the name. Lorenzini speculated that they may be chemoreceptive, but Atema's mentor in the Netherlands, a man named Adrianus Kalmijn, demonstrated their electrical function in the 1960s. At the same time, Kalmijn, Atema, and other students determined the sensitivity of the shark's lateral line to water motion. Sharks typically have thousands of electro pores, a number that remains fixed throughout the shark's life.[3] The scalloped hammerhead shark, for example, has more than three thousand electro pores. The exact arrangement of the canals varies from shark species to shark species.

Part of the excitement about these discoveries is that the new information is overthrowing old assumptions about sharks. For instance, it was once believed that when an animal's muscle contracted, it created a low-power electrical field, which the shark picked up. Scientists then discovered that all living tissue generates a tiny electrical field in seawater. The electrical resistance of an object is a measure of its opposition to the flow of the electrical current. Because the skin of a fish or a sea lion offers substantial electrical resistance, the internal electrical field flows out of the mouth and anus, two low-resistance areas, creating a dipole field that sharks can detect, though that field is minuscule. Muscles and nerves also generate electrical fields, but these are not the main source of detection.[4] A shark's ability to pick up extremely low voltages is extraordinary: the threshold of sensitivity is as low as 15 billionths of a volt.[5] A shark, whose tissue is only flesh and blood, matches humankind's most technologically advanced instruments for detecting electric fields. Injured or sick fish produce stronger electrical fields, but even these fields are faint. Besides predation, this system is also believed to help sharks navigate by sensing the earth's electromagnetic fields.

A shark's bite must be exquisitely timed, and the ampullae are key to this timing. Many attacks between hunter and hunted end in failure. If the bite is off a millisecond, the predator will miss its prey. The ampullae, which are only effective at distances of 3 feet or less, are used in the final stages of the attack. Atema's 2014 paper demonstrated that, regardless of species, virtually all sharks hunt the same way—from the relatively small dogfish shark to the mako, one of the top apex hunters in the ocean.

Spanning the globe, makos are found everywhere, from Montauk to Mumbai, patrolling tropical and temperate waters alike. The mako is one of the most remarkable and unique sharks in the ocean. The word "mako" is a Maori word, though no one can say with any certainty what it means. Though many believe it's a corruption of "mackerel," it's more likely simply the name for the shark. The mako has a dramatic contrast between the vibrant ultramarine color on its dorsal surface and its snowy white underbelly, which camouflages the mako from prey above and below. Between the two, alongside the shark's body, runs a band of reflective silver.

The mako's countenance is severe, and its eyes are black with purpose. Its teeth, which flash like a dagger upon coming into view, are not serrated like the teeth of the great white and other sharks. They are designed to grab and hold prey, not to tear flesh into pieces. Sharks do not chew their food. The teeth of most sharks are slanted inward when the mouth is closed, but they straighten when the jaw opens. The mako's teeth, however, remain upright in the bottom jaw, readying them for action.

Three characteristics define mako sharks: intelligence, speed, and toughness. As a species, a mako is basically a fearless, seagoing linebacker on a Fulbright scholarship.

The mako has one of the largest brain-to-body ratios of all the fish in the ocean, which has prompted several international tests to determine a mako's intelligence and ability to identify objects. A New Zealand scientist demonstrated that the mako can differentiate items and recognize shapes. Perhaps more telling, the sharks involved in the study, after initial caution, started to recognize the scientists as nonthreatening coinhabitants of the water. They allowed scientists to restrain them briefly, and even touch them.[6]

A second defining characteristic of the mako is its speed. Makos can grow to weigh more than 1,000 pounds. Despite its significant weight, the mako is the fastest shark in the ocean. Researchers have clocked the great white at 30 miles per hour, but the shortfin mako can achieve extended bursts of 45 miles per hour. Some people claim makos can reach 60 miles per hour, although scientists consider these estimates excessive. (While the mako beats out all other sharks, the sailfish is the fastest fish in the ocean, clocking 65 miles per hour, the speed limit on the New Jersey Turnpike.) The mako's speed is a product of its shape: a perfectly proportioned cylinder that reduces resistance. The mako's head narrows to a single point, creating a sharp snout that helps the shark slice through the water.

The mako also generates speed across long distances because of its unique tail and musculature. Although the tail of every shark species is vertical—as opposed to the horizontally aligned tails of dolphins— each shark features a distinctive tail, or caudal, fin. In most sharks, the caudal fin's upper lobe is longer than the fin's lower lobe. This is true of the thresher shark. The mako, however, is armed with upper and lower lobes equal in length. At the same time, the mako's red muscles, which process the transportation of hemoglobin through red blood cells, run up and down both sides of its body. While red muscles help sharks swim continuously, the location of the mako's red muscles is a significant distinction. Among sharks, only the great white shares a similar location of red muscles. Like a strong core in humans, the red muscles help steady the body, limiting wasted lateral movements. Various motion studies show that the body of a mako remains fairly stable when it swims. This lack of lateral movement preserves kinetic energy. The mako uses this excess energy to swing the equal lobes of its caudal tail side to side in wider, more efficient arcs than other species. Together, the mako's symmetrical caudal fin and the location of its red muscles allow the mako to generate and sustain such high speeds over long distances.

Another boost to speed comes from the mako's massive gills, which, if laid out flat, would cover a surface area of 56 square feet, and are nearly three times larger than the gills of a blue shark. These massive gills filter and process large quantities of oxygen-rich seawater to power the muscles. And, because makos are endothermic, they can increase their body temperature—sometimes to as much as 20°F warmer than the surrounding water. Thanks to an internal network of blood vessels, the warm blood coming from their large swimming muscles symbiotically transfer heat to the cold blood operating the gills, which in turn delivers some much-needed oxygen back to the swimming muscles. Unlike most cold-blooded animals, including most fish, which dissipate heat out of the gills, the warm-blooded mako holds on to most of this heat, circulating it back through the body. The additional heat makes their muscles more effective—a 20°F increase in body temperature allows a threefold increase in muscle power[7]—rivaling only great whites in this ability.

Atema pointed out another advantage to being warm-blooded: "It increases sensory perception and brain function." The mako is nearly unmatched in its dominion.

The mako's speed allows it to prey on a wide diet: mackerel, billfish, and even other sharks. However, in deep waters, makos turn to tuna, a much more difficult prey. Catching tuna, which swim like underwater silver bullets through the blue netherworld, presents a unique challenge for makos and, for humans, a fascinating glimpse into the wonders of evolution. To catch tuna, makos had to develop speed, which in turn put pressure on tuna to respond. Like the gazelle and the cheetah, tuna and makos battled back and forth for supremacy and the edge for survival. Evolution worked to develop the adaptations for each species to survive, but instead of diverging, the two converged. Remarkably, tuna and mako morphed into near replicas of each other, as if the gazelle grew the limbs of a cheetah. They have similarly shaped tails, where the upper and lower lobes are the same size. Both have horizontal keels that act like small fins; these are shaped in a way that complements both species' speed with stop-on-adime levels of maneuverability.[8] Both developed the ability to warm their blood.

The third characteristic that defines makos is their toughness. They have the strength for sprints and the stamina to sustain the 6,000-mile round trip across the Atlantic between the United States and Britain. More important, they fight like a prize fighter and have the heart to fight to the death. Hooked by fishermen, makos jump completely out of the water, an act of athleticism rivaled only by great whites. To exit the water, the mako needs a starting velocity of 22 miles per hour. As its speed hits 45 miles an hour, the mako can jump 15 to 20 feet, exceeding Michael Jordan's vertical jump of 48 inches.

The sensors jammed into a mako shark's head resemble the cockpit of the F-35 fighter jet. The mako's sensors are equal in sophistication to the fighter jet's advanced systems except they are bundled in nerves, flesh, and blood.

Combined, these three characteristics—intelligence, speed, and toughness—make makos one of the world's greatest apex predators.

Here is a true story of the battle between a mako and a sea lion, one of the largest mammals in the ocean. The story is informed by Atema's detailed, scientific explanation, based on his decades of studying the underwater-sensing abilities of these sharks.

Off Los Angeles, an angler landed the largest mako ever caught, a record shortfin weighing in at 1,323 pounds and measuring 12 feet long. The angler donated the shark to the National Oceanic and Atmospheric Administration (NOAA), which in turn autopsied it. When NOAA scientists examined its stomach contents, they found in the shark's belly a California sea lion,[9] swallowed whole a week earlier. It takes time to fully digest such prey. The sea lion's head, with its eyeballs, whiskers, and teeth, was still clearly discernible. Its winged flippers and dog-like head were pocked with wounds from the shark's pointed teeth. The shark's stomach juices had eaten away most, but not all, of the sea lion's skin. The seal's rib cage and other bones were still intact, waiting for the shark to slowly dissolve them.

To catch the sea lion, this mako required all of its skill, cunning, and strength. Endowed with incredible strength and world-class speed, sea lions are also armed with razor-sharp teeth and ferocious, powerful jaws. Sea lions can weigh as much as 800 pounds and can travel at speeds of 25 to 30 miles per hour. With a short turning radius and powerful muscles, they can catch up to 50 pounds of fish a day, including species as large and fast as tuna. Sea lions are a worthy match for makos.

Here is a dramatic reconstruction of the battle. This mako usually patrolled far out at sea, but on this day it stayed close to the Los Angeles shoreline for some reason. In the vicinity, a large sea lion, hugging the coast, was making its daily rounds for fish not far from south Los Angeles's Huntington Beach. The sea lion broke the surface for air and dove back down. This sound did not attract the mako. Instead, it was the sea lion's odor, which the mako detected about a third of a mile away. Gliding along the coast, the mako followed the odor plume, cutting west when the sea lion cut west, unaware that the ocean's most deadly and relentless hunter was tracking it for a kill. When the sea lion broke the surface again for a gulp of air, the shark followed, its dorsal fin breaching the surface in the same place for a brief moment before submerging again in pursuit of the sea lion among the shadows of the green kelp beds. Without any visual clues, the mako closed the distance from half a mile to less than a quarter of a mile, using its sense of smell and its lateral lines to track its unsuspecting prey.

After half an hour, the shark finally gained a visual of the sea lion, which meant that the mako was near enough for the sea lion to recognize it. Fear released power, and, fighting for its life, the sea lion soared like a rocket, flipping and undulating to evade the mako. Though the sea lion likely reached speeds as high as 30 miles per hour, the mako was able to keep pace, using its secondary dorsal fin to turn in tandem with the sea lion and pull within 3 feet of it. Like an F-35 pilot with a missile guidance system, the mako's ampullae sensed the sea lion's electrical field, its large brain ablaze with the signals it received from its muscles and pectoral and caudal fins. With one last lunge, the ampullae-directed jaws tore into the sea lion's rib cage. Red blood swirled in the blue water. The sea lion lashed back at the shark and, in a futile attempt to break free, tried to bite the shark with its formidable teeth, though it couldn't penetrate the denticles of the mako's rough skin. Finally, after twenty minutes and a half mile, the mako swallowed the sea lion whole.

In this attack, all the mako's senses came into play, from the moment it sensed the odor to the moment of the final bite. The mako was just too strong and quick for the sea lion, an epic battle between species that has been waging for eons, since the Silurian period—prey and predator battling for survival and evolving over millions of years to create even a small edge that might mean the difference between life and death. In the evolutionary process, nature has created a dazzling array of species, each with its own remarkable adaptations. Equally important, nature maintains an exquisite balance among species, which keeps the system intact, a dynamic equilibrium within an ecosystem in which the diversity and population of animals remain relatively stable. If too many sea lions survive, they feast on too many fish, disrupting the balance of the surrounding ecosystem. The makos maintain this delicate balance as apex predators.

To humans, it may seem cruel for the sea lion to be eaten alive, but that is nature's way. An efficient and quick death is perhaps better than suffering. When the sea lion disappeared into the mako's cavernous maw, it should have been forever, but it wasn't. A week later, another apex predator entered that same area. A sports fisherman, chartering a boat with state-of-the-art electronics, a large bait capacity, and powerful twin diesel engines, let out his fishing line just outside the Los Angeles Harbor and hooked the mako. As soon as the mako clamped down on the hook, it realized its mistake and made a run for it. The reel sang as the line tore through the rod's tip and into the water. The shark took out a quarter of a mile of fishing line. As the struggle continued, the shark, in an attempt to spit out the hook, leaped over 20 feet out of the water, fighting for its life. Over the next hour, as the fisherman struggled to reel in the mako, it executed five leaps.

Finally, after two and a half hours, the fisherman hauled the exhausted mako close to the boat, and one of the crew yanked the shark aboard with a large gaff hook. The crew, captain, and fisherman all gasped in shock at the size of the mako. Large makos are typically 8 or 9 feet long, but this one was 12 feet. Impressed with his catch, the captain took the mako to a special weigh yard, where a certified weighmaster registered it at 1,323 pounds, 102 pounds heavier than the previous world record for a shortfin mako shark caught off the coast of Chatham, Massachusetts.

According to California law, a sports fisherman cannot sell shark meat. He or she can consume it, however, for him- or herself, or donate the shark for research. At the NOAA Southwest Fisheries Science Center, the mako sat on a table, waiting to be autopsied. The specialist came in, sliced the mako to get to the stomach contents, and removed the sea lion from the shark's belly. A week after the epic battle, hunted and hunter both were dead.

Anglers pursue makos because their dreams of an epic battle with the big shark can turn into reality. Makos charge boats, sometimes jumping right into them. In mako fishing, you might lose not only your fish but also your rod or even your arm. This big game fishing carries a real element of personal danger.

For anglers, the mako is the prize shark to catch; the fight it puts up is legendary. History is replete with epic battles like the one between an angler and a mako off Nantucket. The mako flew out of the water and into the boat and whacked the fisherman across the ribs and stomach. Thrown almost the entire length of his 13-foot boat, the fisherman found himself forced against the cabin door, the wind knocked out of him. Gathering himself, he looked down; 8 inches from his left foot was the gaping mouth of a 286-pound mako, its rows of razor-sharp teeth exposed. The shark, it seemed to the stunned fisherman, was eyeballing him, trying to gauge whether he was predator or prey. Suddenly, the shark came to life and, for the most terrifying two to three minutes of the fisherman's life, started chasing him around the cramped cockpit. The fisherman grabbed a baseball bat and struck the mako, which only emboldened the shark. Finally, after six blows, the fisherman killed the mako, though not without first enduring a litany of injuries, including three fractured ribs and a lacerated elbow.[10]

Here is another story from one of the most famous shark fishermen of all time, Frank Mundus, the inspiration for the Quint character in Peter Benchley's *Jaws*.

His story takes place on a fishing trip in Brielle, New Jersey.

As the mako zoomed through the air toward our boat on a parabolic trajectory, he came down squarely across one of the rods. Down the rod slid the shark's bulk shearing off the rod guides as it went. The startled fisherman, seeing a few hundred pounds of mako about to be deposited in his lap, had the presence of mind to get the hell out of there. He jumped out of the fishing chair, going over backward, and landed on his head on the deck, almost breaking his neck. The mako slid into the cockpit and promptly took over.

Pandemonium broke loose. The shark went berserk. Bouncing and twisting violently, he slammed his way from gunnel to gunnel. He smashed fishing chairs; he sent chum cans flying to spew the contents all over the place. All hands beat a hasty retreat to the cabin and barricaded themselves behind the companionway door. Seconds later the mako crashed against the door knocking its knob out by the roots.[11]

In my travels, I heard a more recent—and much more sinister—take on a familiar tale. In 2017, three men—Michael Wenzel, 21, Robert Lee Benac, 28, and Spencer Heintz, 23—gassed up a high-speed boat and headed out to fish. Along with a case of beer, they took along a revolver and 50 feet of rope, hell-bent on catching a mako in Tampa Bay.

It wasn't long before the three men caught a mako. As they reeled it in, they brought it alongside the boat. Exhausted after its struggle, the shark was moving slowly, like a punch-drunk boxer in the fifteenth round. Wenzel went over to the boat's helm and removed the gun from a small cabinet. Aiming the muzzle at the shark, he pulled the trigger. A bullet ripped through the shark's sandpaper-like skin, lodging in its liver. Because sharks are powerful animals, a single gunshot isn't always enough to kill them, at least not right away, which gave the men some time for "fun." It was time to teach this shark a lesson.

Next, they pulled out the rope and tied one end around the tail of the shark and the other around a cleat at the stern. Wenzel returned to the helm and gunned the engine, dragging the shark at 40 knots in the wake of the boat for more than five minutes.

The three men took a video of the ordeal. In the video, Wenzel, Benac, and Heintz are laughing and smiling as the shark struggles in the water. The ordeal and gunshot wound are too much for the shark, and it ultimately dies under this torture. The men hold up what is left of the carcass to the camera: a heap of blood and tissue and, because most of the shark's skin was torn apart, exposed cartilage.

The men were proud of what they had done. To show off, they sent the video to Mark Quartiano, a legendary charter fisherman in Miami known as Mark the Shark. Quartiano takes people out specifically to catch sharks, and by his own account he has killed hundreds of them. Yet after viewing the video, he was filled with outrage. "I was horrified," Quartiano told CNN. "I've been shark fishing for fifty years, and I've never seen a disrespect for an animal my entire career that was that evil."[12] Quartiano said he decided to publish the video to Instagram, so the fishermen could be identified and caught. Under the hashtags #sowrong and #notcool, the video racked up thousands of views. It attracted the attention of Florida wildlife officials, who began an investigation. Working with the Florida Fish and Wildlife Conservation Commission, the Hillsborough County State Attorney's Office charged two of the men, Wenzel and Benac, with two separate charges of aggravated animal cruelty, a third-degree felony, and a misdemeanor charge of using an illegal method to take a shark. In his police photo, Wenzel is smiling.[13] Florida governor Rick Scott said he wanted laws to prevent "such inhumane acts."[14] (As of this writing, there has been no court decision regarding guilt or innocence.) The men who tortured the shark were shocked that Quartiano had, in their eyes, betrayed them. In disgust, they hurled at him what they considered the ultimate insult; they called him "a PETA-lover."

I met Mark the Shark in his office, which is located on the banks of Biscayne Bay, directly across from downtown Miami. Dangling from the ceiling were scores of shark jaws, turned white with age and filled with multiple rows of enormous jagged teeth. Jaws more than a couple of feet wide meant those sharks must have been 20 feet long, if not more.

Quartiano is a muscular, powerfully built man with blond hair and blue eyes. In a past life, he could have been a Viking raider, or in a more recent one, a WWE wrestler. He considers his boat the best office in the world. We talked about his business and, as he describes it, his role as the "last shark hunter on the planet." He told me, "Some charter guys don't even want to go shark fishing anymore because there's just so many restrictions on these fish. Doesn't even pay when they get the fuel and they got their boat and they can't catch a fish and keep it. It doesn't make any sense for some of these guys."

Quartiano views his job as being to catch the largest

fish in the ocean for clients who come from all over the world.

"We take out every kind of person," he said. "Right now, we're taking out a lot of kids. The new generation is fascinated with sharks. And we can let them go or tag them, and it's an experience these kids will never forget." He added, "Then other clients are the trophy hunters who yearn to catch a big shark—a hammerhead, a big tiger, or a big thresher shark. Something big."

Due to the Gulf Stream, the Miami area is a highly productive fishing ground, which means it's also a rich territory for sharks. The Gulf Stream, a warm water current, originates in the Caribbean and travels all the way to Europe, distributing the warmth from the equator to the northern reaches of the planet. The Gulf Stream creates the mild, wet winter for Europe, and without it, continental winters would be much colder. The Gulf Stream also acts like a giant highway for fish traveling from the Caribbean up to the Gulf of Maine. Some species of Caribbean fish, like triggerfish and mahi-mahi, inadvertently get sucked into the current and end up off the coast of Massachusetts and New York, falling victim to the Atlantic's icy waters in winter. In addition to the Gulf Stream, Miami possesses some of the best reefs in the world, which are situated only a mile or two from shore. The twin presence of large reefs and the Gulf Stream attracts big game fish. This cornucopia of trophy fish like sharks brings in fishermen from all over the country.

When I asked him about his reputation, he shook his head. "Mainly, a lot of people think that I'm just going out there and killing everything that swims, and [that's not the] case at all. I mean, we catch and tag so many fish all year long. Nobody wants to hear that part."

Clients make the ultimate decision about the fate of the shark, according to Quartiano. "They're the ones that say, 'Mark, I want to kill that fish. I want to put that fish and those jaws on my wall.' A lot of groups do put pressure on me, as far as the catch and release. But again, a lot of these fish aren't going to survive when you catch them. Unfortunately, they're gonna die on the line. So, really, if you care about these animals, you just leave them alone."

I asked him what has happened to the shark population over his career. "The commercial fishermen definitely have had an impact on the shark population over the last twenty years or so," he told me. "We used to guarantee our customers a shark; otherwise, they wouldn't pay. Now, because of all the pressure on the sharks in this area, we can't do that."

Quartiano doesn't think putting pressure on recreational fishermen to cut back their fishing is going to replenish the shark population. "It's the commercial guys that wipe out just about everything, with their indiscriminate killing of everything. The bycatch is enormous and is just tossed back into the sea. Stop or just cut down the commercial fishing on some of these big game fish and we'll be fine."

I did some fact checking on Mark the Shark's statement, and while statistics are not readily available for recreational shark fishing, it can have a significant impact on fish populations and sometimes even more so than commercial fishing; in 2013, the annual landings of sharks in recreational fishing (4.5 million pounds) exceeded that of the commercial sector (3 million pounds).[15] One research paper studying the impact of recreational fishing on sharks concluded that this trophy-hunting activity presents a growing risk to shark populations.[16] In spite of this concern, as long as the killing of sharks is romanticized and celebrated, Mark the Shark will continue to have a business, and anglers around the world will continue to make sport of hunting sharks, a sanctioned pastime that kills thousands of sharks every year.

Some conservationists in California were dismayed when the record-breaking 1,323-pound mako was destroyed instead of released. Unlike great whites, makos

EMPERORS OF THE DEEP · 81

are considered "endangered" on the IUCN's Red List. One reason they are endangered is that makos need an extended time to reach sexual maturity. Female makos reach sexual maturity at age eighteen. By this time they have usually grown to 10 feet in length, big enough to carry shark pups over their fifteen-to-eighteen-month gestation period, after which the female gives birth to a litter of four to twenty-five offspring.

Another reason the population grows slowly is that these animals do not breed that often. Males and females converge to mate only between late summer and early fall. Female makos wait about eighteen months after giving birth to get pregnant again because the pregnancy is so stressful. Therefore, females only reproduce every two or three years. As a result of this breeding pattern and overfishing, the species has been struggling. They often end up trapped in gill nets, snatched up by longlines as bycatch, and killed in shark tournaments like the one I witnessed in Montauk, which places further pressure on the shark.

Because I wanted to understand the fascination with killing sharks for sport, I revisited a mako-only shark tournament in Montauk, the Super Bowl of shark contests. There I met a kindly gentleman I'll call Tom, a longtime veteran of these tournaments. Tom was in his early sixties and had been a fisherman almost his whole life. He explained how the tournaments operate. Fishermen pay an entry fee to participate, usually $975 to $1,000. Participation doesn't require qualification or technical skill, just access to a boat and some fishing tackle. Fishermen vie against one another in different categories, including species and size: largest shark overall, the largest mako, the largest thresher, and so on through the water's numerous shark species. The winner of each category receives a trophy and prize money, which the tournament director presents at the end of the tournament. "The fishermen aren't just interested in catching a shark," Tom told me. "They want to spice things up, and the way to do that is to add an extra element of fun."

And there's no better way to have fun than gambling, according to Tom. The night before the competition begins, fishermen gather for a celebratory dinner, typically at one of the restaurants along Montauk's marina, where they wager on the weekend's outcome. Rounds of drinks loosen lips, and boasting loosens the fishermen's wallets, kicking off a betting frenzy. While different types of betting are popular at tournaments up and down the Atlantic, the preferred game here is the Calcutta auction, in which the participants auction off the boats. Whoever bids the highest for a boat then "owns" it throughout the weekend tournament. If that boat catches the largest shark, or the largest shark in a specific category, the "owner" is entitled to a portion of the prize money and the betting pool. "Someone can buy their own boat or someone else's boat," Tom said. The boat owners can each wager anywhere from $5,000 to $25,000. Adding up all the bets from all the boats means that the prize money can total half a million dollars or more.

With that much money at stake, the temptation to cheat is great. The participants realize that some extra help can earn them tens of thousands of dollars. One way some cheat is to add lead weights inside the fish. Another 50 pounds on the scale can tip the shark into the winner's circle. This approach is risky because it is easy to spot, so some fishermen resort to other measures, such as adding water into the guts of the fish. Water weighs over 8 pounds per gallon, so this approach can give a dishonest team that crucial edge. Unsuspecting judges can easily miss this, but the more seasoned ones will make sure that all excess water has been drained from the shark's body cavity. Another nefarious trick is to purchase a large shark from a commercial trawler, which are prohibited from entering the tournament but often have unwanted sharks.

The tournament has a time limit: the teams have to be back at the dock at 6 p.m. unless they radio in that they are at the inlet. The more time out, the better the odds of catching something, so some teams will radio at 6 p.m. that they are at the inlet when they are, in reality, still far at sea in the prime fishing grounds. Then they will run the engines flat out to get back within a reasonable time.

To experience the tournament firsthand—and to film it as part of a documentary I was making about shark hunting—I hitched a ride on one of the competing boats. Shortly after six in the morning, my camera crew and I arrived at the dock. It was a gorgeous June morning, the sun already bright above the horizon, and the sky had a touch of pink against the bottom of the cumulus clouds above. The boat's captain, whom I'll call Stan, welcomed us aboard. A happy, heavyset man in his fifties, Stan sported a neatly trimmed goatee and spoke in a gravelly voice that reminded me of Robert De Niro's character in *Raging Bull.* Showing us around the boat, he fingered a cigarette and drank regularly from a blue Solo cup, which was filled with an unidentified liquor. Unsolicited, he showed me the battle wounds on his hands. "Tore up this finger with a hook," he said. A chunk of his thumb was missing. "Took another hook here," he noted, pointing to a jagged scar on his wrist.

Besides the occupational hazards, I asked Stan how

he was feeling. "Lucky," he said. "Better to be lucky than good. Today's our day."

The six members of Stan's crew must have felt the same way. Despite the early hour, they were prepping the boat with enthusiasm, unconcerned with the cameras and the men carrying them. While I fastened a life jacket, I noticed none of his crew bothered to put one on. Sam, an older man in his late fifties, loaded Igloo coolers with ice and bluefish, which would later be used as chum for the sharks. With the weather-beaten face of a lifetime fisherman, he packed the containers like an expert, as casually and efficiently as if he were readying a backyard barbecue.

The tournament director assigned me to this crew and boat, a purpose-built, top-of-the-line charter, which, at 60 feet long with an 18-foot-wide beam, was clearly designed for luxurious fishing. The spacious cabin had leather-padded chairs and other amenities: a wet bar, plush carpeting, and a refrigerator, fully stocked. Near the stern, a wide soft sofa in the main cabin invited guests to gaze out over the ocean, eating and drinking below board, while the fishermen hunted sharks.

We embarked from Montauk Harbor and moved past the town's iconic lighthouse. The stern sent out a wide wake behind us, and a short twenty minutes later, we were out of sight of land and in the choppy, blue waters of Long Island Sound. Close to fifty other boats jockeyed for position in the open sea. Stan cut the engine and settled the boat in one of the most productive fishing grounds in the world, where he suspected makos congregate. Because warm waters bring fish, makos track the Gulf Stream up the East Coast, from Montauk to the Gulf of Maine. The fish that live in Long Island Sound will eventually mature and migrate into the Atlantic Ocean off Montauk.

Sharks have no interest in these boats, nor the usual bait of squid or fish parts. To attract the sharks, live bait was needed to supplement the chum. Stan's crew took out fishing rods and quickly caught two bluefish. They hauled the fish on board and tossed them into a live bait well. Sam held down the fish, while Stan jammed a fishing line through the bluefish's heads, just above their eyes. He attached the line to a loop and threw it back in the water. The blues swished their tails to escape, but the fishing line held them. Assigned the task of making chum, Sam put the bluefish from the Igloo into an electric meat grinder. Blood and fish parts oozed out and fell into a white bucket. He emptied the bucket's contents overboard, creating a slick of red across the blue ocean. The current took the scent out past the boat into the sea.

His task completed, Sam set down the bucket and kicked back, waiting for the sharks to show. "Now is the time to drift and dream," he said.

As the boat bobbed in the water, I took the opportunity to talk to Stan's twenty-five-year-old son, Barry, who grew up fishing. "I've been doing this pretty much my whole life," he said, punctuating our conversation with a cigarette. "Sometimes you expect one thing to come out of the water, and something else entirely comes up. There's always that unknown. It doesn't get old."

Another half hour passed, and nothing happened, then an hour. Still nothing. We were staring down sheer boredom. The crew puttered around, throwing more chum out into the water. Some men drank beers to cool off and pass the time. Sam turned on the radio, drowning out the sound of the waves lapping on the sides of the boat. As I stared out over the horizon, I wondered how long I was going to be stuck out there. Then, far beneath the waves, a mako picked up a scent and decided to veer off course and investigate. The fishing pole started to twitch, just a slight bend at the tip of the rod. The rod stopped twitching and straightened. But it started again, a little more bend in the rod each time. A shark was nibbling at the bait, testing to see if there would be any reaction and what kind of resistance it might encounter. And then the shark decided to strike. Suddenly, the line ripped out of the reel and charged down the length of the rod into the sea. The shark was making a run for it. Sam grabbed the rod and placed it in a holder in the captain's chair at the stern of the boat. Barry hopped into the captain's chair, strapping himself in for what proved to be a lengthy, exhausting battle. The mako shark is built for speed, and its muscles powered its tail like a modern diesel train.

Barry started reeling in the line, but every so often the fish decided to fight back and make another lunge out to sea. Barry could only watch as the line he had struggled to bring in darted back out. This process—Barry reeling in the line, the mako taking it back out—carried on for half an hour, then an hour, then two hours. Barry's gray T-shirt was drenched with sweat, and his face showed the strain of his struggle. The strength of the shark had taken its toll on him. At that point, he stopped reeling to give his tired hands and forearms a break. Staring at the sky, he took a deep breath, then continued reeling in the line. I was standing a few feet behind him, amazed at Barry's determination, and the shark's. Anglers live to catch makos. "They fight hard," Barry said. Like clockwork, the shark jumped out of the water about 20 feet in the air. The entire crew cheered, but this wasn't a game for the mako; it was life or death for the shark. To my amazement, the shark jumped again, achieving the same height as its previous jump. I counted five total jumps, a feat that required amazing strength and will.

Realizing this standstill could continue for another couple of hours, Stan started the engines. When I asked him what he was doing, he said, "If the mountain will not come to Muhammad, then Muhammad must go to the mountain." Thick black smoke poured out of the exhaust pipes at the stern of the boat. The engine noise drowned out the radio, and yard by yard, the boat backed up methodically toward the shark's position, closing in on the kill. I couldn't help but think they were cheating, because two hours into the fight, the mako was no match for the boat's twin 1,200-horse-power engines.

Barry unstrapped himself and stood up next to the gunwale. He frantically reeled the line. I peered into the frothy mix of blue and white water and made out a shadow. Barry saw it, too, and strained to get the shark to the side of the boat. Stan pulled out a gaff, a large stainless-steel pole with half a harpoon on the end. He gashed the shark in the gills. Trying to drag it out of the water, he yelled, "C'mon, you sonuvabitch!" The shark was now clearly visible. It rolled over, exposing the contrast between its dorsal-side marine blue and the white of its underbelly. Just below the water's surface, the head of the shark appeared, and with a final yank of the gaff, Stan managed to pull the mako on board. I could see a part of the baitfish sticking out of the shark's mouth. The shark thrashed around, moving its head frantically from side to side. Its sandpaper skin scratched the floor. It sounded like a saw trying to tear apart the wood. Someone—in the melee, I couldn't tell who—yelled, "Watch your feet!" Blood splashed over the floor of the stern. I had to turn away. I couldn't watch the shark suffer. I stared at the opposite horizon, but I couldn't silence the violent sounds of the shark scratching against the floorboards.

Sharks can survive for a long time out of the water, and as I discovered, it is difficult to tell when the shark has died. Anyone presuming a shark is dead runs the risk of getting bit. To protect the passengers, or perhaps to expedite the process, Barry unsheathed a knife. Though I didn't want to, I returned my attention to the shark. Barry straddled the fish, positioning himself right behind the shark's gills, far enough away from the shark's mouth that it couldn't turn and bite him. He was surprisingly calm as he bent over and entered the knife deep into the mako's anterior, or front, end. Deeper and deeper he sank the knife until Barry felt the resistance of the spine. That's when, to my horror, he started sawing back and forth, severing the spinal cord. Blood oozed out of the incision, and the shark twitched, its life slowly ebbing away. Barry stood back, blood smeared all over his fishing wader, and we all waited a few minutes, watching the lifeless shark remain still on the deck, until we understood it was finally, mercifully, dead.

To beat the 6 p.m. deadline, Sam brought us back to port. On our return to Montauk, I asked Stan what he thought about the shark, which remained in the stern, untouched. Shrugging, he said, "A shark is a shark. Something is better than nothing."

The boat finally arrived at the marina, where Stan and the crew dropped off the shark. A crowd of onlookers waited near the slip, like spectators in the Roman Colosseum, to catch a glimpse of the shark's bloody, mangled body. One of the dock hands threw a line with a loop at its end so the fishermen could slip the tail of the shark through it. Conquered, the fish could now be hauled up onto the yardarm and weighed. As the winch pulled the line and the shark went up the yardarm, the crowd murmured at the sight. "One hundred and fifty pounds," the announcer called out to the crowd. It was a young shark, judging by its size and weight, too young to have ever bred. Because the minimum weight requirement for the tournament was 175 pounds, the shark was disqualified. It died for nothing.

My camera crew and I got off the boat. I watched Barry light another cigarette and Stan refill his blue Solo cup. Once on the dock, Stan took the mako's carcass and cut off the tail. He drilled the mako's tail onto the pier as a trophy. As I toured the marina, I witnessed other fishermen cut their hauls into pieces. One man chopped up a thresher shark and stuffed its decapitated head into a blue Igloo cooler. The fisherman told me he wanted to take the head home with him to carve out the jaws as a trophy. In the meantime, he draped the thresher's tail over his shoulders, like a shawl, while his buddy snapped a picture. Other men loaded their shark heads into a 4-foot-long gray cart, which made it easier to transport these gruesome trophies to their vehicles in the marina's parking lot.

The entire dock was a macabre Dantesque scene in which mako, blue, and thresher sharks were in various stages of dissection. Off to the side of the square waited a tractor with a large front loader. Men, whose dark slickers were wet with shark blood and viscera, filled the loader with shark parts. One man brought over a shark heart, still beating, for the crowd to see.

Some sharks were dangling from the yardarm,

blood running down their bodies. Over the next hour, the announcer continued to call out the weight of each shark. Another mako was hoisted on the yardarm. It weighed 356 pounds, which drew a collective gasp from the crowd. It was a few pounds off the leading fish, though, so the captain later butchered it as a visiting family looked on.

The front loader was brimming with heads, fins, intestines, and other body parts. An operator fired up the tractor, and it crept away from the crowds into the shadows of the marina, where these shark parts were tossed into a dumpster.

People were having drinks and eating just a few yards away from the carnage. Some milled around the marina, gazing at the sharks strung up on the yardarm. Children holding on to their mothers' hands walked in between the shark carcasses. I asked a young mother what she thought about shark tournaments. "It's great," she told me, "the whole family can get together."

The young fisherman who caught the largest mako walked away with $90,000, and Tom later told me another fisherman took home $250,000.

While some humans benefit from the gambling, trophy fishing can have a negative impact on the shark population. Tournaments target the biggest and strongest sharks, and when these are lost, the event has a disproportionately large effect on a species's ability to sustain itself or recover from overfishing and climate change. In other words, the loss of the strongest individuals hurts the gene pool of the shark population even when relatively few individuals are removed.[17]

Currently, there are approximately seventy official shark tournaments up and down the Atlantic coast, from Maine to Florida, twenty-eight of which target individual shark species. New York State allows these tournaments to take place, and thirty-seven different local businesses and multinational corporations sponsor them. Following the tournament, as I tried to wrap my head around how and why this kind of sanctioned brutality is allowed to continue, I remembered that another tournament was scheduled in Montauk four weeks later. This carnage, championed as sport, would be repeated, an ongoing but totally preventable surrealistic nightmare.

Chapter 3
The Mysterious Case of the Hammerhead

Frank Fish, PhD, was perplexed. Looking at a sculpture of a humpback whale, he noticed a series of bumps on the forward edges of the whale's flippers. As a professor of biology at West Chester University of Pennsylvania, he is a widely regarded expert in the dynamics of locomotion in aquatic mammals, but he didn't know why those bumps were there. (Dr. Fish enjoys the irony of his name.) He understands well that nature is far from frivolous with its designs. Those bumps were there for a reason; he just didn't know why. In 2007, after confirming that the artist's rendition was accurate, Fish set about trying to figure out what purpose the bumps serve.

Backed by a grant from the National Science Foundation, he and a small team investigated the whale's pectoral flippers, gigantic 15-foot-long wings equal to one-third of the whale's body. These wings, they discovered, serve multiple purposes. Aiding the giant whale's ability to turn laterally, or side to side, pectoral flippers also allow the whale to swim up and down, increase stall speed, and perform a host of other maneuvers that allow the whale to thrive in the oceans. To measure how these bumps work, Fish collected flippers from deceased humpbacks and took a computed tomography scan (CT scan) of them, which he then used to create a three-dimensional model of the flippers, bumps included. Next, Fish placed the models in a water tunnel. Measuring lift and drag, he adjusted and tested the model at various angles in different currents. His study—the first hydrodynamic analysis of a whale's evolutionary design—proved that the curious bumps he first spotted on a sculptor's rendition of a humpback greatly enhanced the whale's ability to move freely and quickly underwater, despite its massive size. Performing rolls and loops, swimming 14,000 miles per year—all are possible thanks to the bumps. And when the 30-metric-ton leviathan wants to soar out of the water, its bumpy flippers help it to do so. Fish called the bumps "tubercles" after the Latin word for "protuberance," and he quickly realized that these previously unknown bumps greatly improve both hydrodynamic and aerodynamic performance. Fish started integrating tubercles into energy-efficient wind-turbine and fan blade design at WhalePower, his Toronto-based, for-profit venture.

I met with Fish at the American Museum of Natural History in New York, where we discussed the scalloped hammerhead shark, which like the humpback also has tubercles. "Tubercles evolved as a way to improve pitch, or up and down, performance," Fish told me. "They add more lift, and conversely, they can also help the shark when it is pitching down." Such freedom of movement lets the hammerhead hunt and catch prey on a par with its fellow Big Four cartilaginous kin. More than any other shark, though, questions circle around the hammerhead, unanswerable mysteries that extend far beyond the shark's trademark T-shaped head, called the "cephalofoil." Getting to know the hammerhead, I quickly learned, brings one into contact with the mysterious.

In 2010, curious about how hammerheads evolved over time, scientists at the University of Colorado at Boulder set about to construct the shark's phylogenetic, or evolutionary, family tree. They collected mitochondrial and nuclear DNA from dead hammerheads at fish markets around the world. What they found is that all nine species of hammerheads share the same ancestral father, literally the same massive shark. The study's lead author, Andrew Martin, determined that this 20-foot-long behemoth broke away from other shark species about 20 million years ago, undergoing a divergent evolution during the Miocene epoch, when the global climate was warmer.[1] Some new species branched out into different developmental directions, separating into today's nine hammerhead species, which vary in size and behavior and are found all over the world. The great hammerhead shark, the largest and most recognizable of the nine, kept the size of its ancestor. One of the larger sharks in the ocean, great hammerheads grow up to 20 feet in length and, on average, can weigh more than 1,000 pounds. Most of the other eight hammerhead species measure in at much smaller sizes, a process known as neoteny.[2] In neoteny, adult sharks retain juvenile traits, and the species invests its energies in reproducing instead of growth. For instance, the tiny bonnethead, which looks like a great hammerhead doll, maxes out at 2 or 3 feet in length. Despite the variety in size and behavior, all nine species feature brownish-gray-and-white coloring on their underside.

Unlike any other fish in the ocean, the hammerhead surveys its underwater environment with its trademark head, which is crammed full of flesh-and-blood sensors. Many have wondered why its head is T-shaped.

One theory is that the shape of the head improves the animal's vision. Perched at the extreme ends of its head, a hammerhead's eyes afford it a 360-degree panoramic view of its environs. Michelle McComb of Florida Atlantic University planted electrodes into the eyes of three hammerheads—scalloped hammerhead, bonnethead, and winghead—and moved a light beam across them to measure the eyes' field of vision. She found that the degree of overlap between the eyes increased as the head of the hammerhead species widened. The wider the head, the better the depth perception. Previously people had thought that the T-shaped head was to improve the shark's sense of smell, but McComb's research shows that the shape of the head improves the shark's vision.[3]

And it's not just its eyes that are unique. In addition to its unique ability to survey its environs, the hammerhead has two noses. Its double-barreled olfactory system enhances the shark's ability to assess what's going on around it. Jonathan Cox, a senior lecturer at the University of Bath and a noted expert in fish olfaction, has been working to understand how the olfactory organs work inside a hammerhead. Using a scale model of a hammerhead shark in a flow tank, he figured out how water flows through the shark's nasal cavity. The model, Cox told ScienceDaily, was complete with internal cavities accurate to less than eight-thousandths of an inch. He and his team submerged the three-dimensional replica into a flow tank and observed the flow of water. "Whereas humans use their lungs like a bellows to inhale air through their noses to smell, the hammerhead shark smells as it swims forwards, propelling water through its nose," he said. To simulate the shark's style of swimming, he went on, "we change the angle of the head model in the tank and observe the flow at each angle."[4]

A hammerhead's nasal cavity, which is located at the outer edge of the cephalofoil, is a labyrinth of pipes, with a central U-shaped channel and several smaller channels leading off it. The smaller channels contain the olfactory receptors. As the water flows through these channels, the hammerhead is able to sample the water for telltale odors. When the hammerhead swims forward, it sweeps its head side to side. Picking up on an odor plume, the shark can use the scent's different arrival times in each nose to track it to its source. Because the distance between the shark's polar noses is significant, the arrival times of the odor plume are staggered, which lets the hammerhead more accurately figure out where the scent is coming from. Like other sharks, hammerheads have embedded-in-the-head ampullae of Lorenzini, the electroreceptor sensory pores that detect electrical disturbances in the water. Coupled with their well-spaced noses, the similarly spread-out ampullae improve the hammerhead's ability to sweep for prey effectively, detecting creatures buried in the sand. They do not have to see them to know that they are there.

Another distinctive aspect of the hammerhead is its dorsal fin, which is ramrod straight and can reach a height of almost 2 feet. A shark's dorsal fin acts like a rudder on a boat to stabilize the shark's turn. However, in the case of the hammerhead, the dorsal fin is longer than the pectoral fin. So the hammerhead swims sideways to take advantage of the lift from the dorsal fin, like a wing. The scientists at the University of Roehampton in London tracked several great hammerheads and noted that 90 percent of the time they were swimming at roll angles of between 50 and 75 degrees. When the group performed wind tunnel studies, they found that sharks would use 10 percent less energy when swimming sideways.[5]

The hammerhead's arc-shaped mouth lies on the underside of the head and is small compared to that of other sharks. Some underwater photographers relish taking relatively close-up photos of hammerheads because they know no real danger exists. But make no mistake: the hammerhead is a masterful apex hunter, one that employs all of its highly evolved equipment in pursuit of prey.

Hammerheads and stingrays, the hunter and the hunted. Over millions of years, the drama between these two arch enemies has played out, a competition every bit as intense as the legendary contest between the lion and the gazelle, or the mako and the tuna. The hammerhead evolved into a supreme hunter capable of detecting stingrays invisible to the naked eye. Yet the stingray is a formidable adversary. One stingray, the spotted eagle ray, can race fast enough underwater to reach the speed necessary to fly completely out of the water. Once the ray is airborne, flapping wings replete with dripping water, it can sometimes fly over enough distance to escape the hammerhead. If the stingray can't escape the shark, then it can stand and fight using its deadly stinger, a hard, thin spear filled with venom located a foot from the tail's base. If something comes too close, the ray will fling its stinger into the victim. If a human is stung by a stingray, it is painful but usually not fatal unless the victim is stung in the chest or abdomen.[6] When hunting a stingray, a hammerhead must work quickly to neutralize the stinger. A skillful hammerhead can disable the stingray by biting its wings, completely avoiding the stinger. However, many sharks do catch a stinger to the mouth. Some hammerheads have been spotted with more than a dozen barbs around the mouth.

In the Bahamas, underwater photographers have captured hammerheads in action as they prowl along the sandy ocean bottom for stingrays. One dramatic chase was captured on camera near Bimini, which has some of the best visibility in the world for scuba diving. Here the visibility is 100 feet, which is the equivalent of seeing an entire baseball diamond underwater. The life-and-death chase started with a hammerhead shark patrolling in 25 feet of clear turquoise water. The hammerhead swam less than a foot above the seafloor, which was pure white sand pushed into neat furrows by the tides. Occasionally his pectoral fin inadvertently clipped a bit of furrowed white sand as he swept the area. His head and body twisted side to side, and his eyes flashed back and forth on the lookout. His head scanned the bottom as he moved forward, and with all this information, particularly the electrical signals from the seafloor, his senses told him a creature was buried in the sand. To human eyes, there was nothing here. His body twitched and jerked with anticipation. The shark started to shrink his search area, and the circle he was making began to tighten. With each sweep of his head, like a radar antenna, the shark divined the exact location of the hidden stingray. He focused on one area. The ray could sense it was only a matter of time before he was discovered. He unfurled his wings and, with all his might, pushed out from the sand, leaving a small cloud behind him. The chase was on. The hammerhead had flushed out his prey, and the stingray hoped to outrun him. The stingray surged ahead, twisting and turning to throw off his attacker, but the hammerhead matched every move and slowly gained ground. He drove down on top of the sting-ray and used his head to pin his prey to the bottom and immobilize him. The stingray squeezed out from under the head and spurted again into the open. The hammerhead flicked his tail and, with his dorsal fin providing stabilization, again closed the gap. At last, he lunged and took a bite out of the wing, sending the stingray into shock. The hammerhead began feeding on his prey. He was fortunate in that he had avoided the stinger. But in the underwater world, commotion brings attention—many times, unwanted attention, and there was something else watching the hammerhead besides the cameraman. A bull shark was watching the proceedings, and he wanted a piece of the prize. The hammerhead took off with the stingray in his mouth and the bull shark followed, nipping at the remains of the stingray hanging out of the hammerhead's mouth.

Both disappeared from the camera's view into the misty turquoise waters.

Science has uncovered how hammerheads hunt, but that is just one part of their lives. I often wonder about the daily lives of hammerheads: Do they sleep? What do they do during the day? And where do they travel? Research has emerged revealing the why and where of hammerhead behavior beyond feeding.

Hammerheads are found throughout the world's oceans, although most hammerhead species prefer warm waters, usually congregating along the coastline in shallow water. Not the great hammerhead shark, though, which lingers in deep water. Scientists have known for a while that hammerheads migrate in the summer to warmer waters, but until recently the details of their migratory patterns remained a bit sketchy. In 2006, a multi-institutional group of scientists affiliated with the Charles Darwin Foundation, Galápagos National Park, and the University of California, Davis, conducted a four-year study in the Galápagos. Charles Darwin came to this archipelago to study the development of animals, which led to his theory of evolution. He found the islands depressing and desolate, inhospitably dry, with little vegetation. But because the islands are scattered among 200 miles in the ocean, species-specific adaptations occurred in isolation. New species could develop based on adaptations necessary for survival. A mix of cold water from the south and warm currents coming from the equator brings abundant sea life to the Galápagos, an obvious draw for sharks. Being located 350 miles off the coast of Ecuador helps, too, because the islands are far enough away from the mainland for sharks to swim freely, unmolested by humans. This makes the islands the ideal place to study the behavior of hammerheads.[7]

In the study, scientists caught and tagged with special acoustic and satellite telemetry devices more than 130 sharks in the northern Wolf and Darwin Islands in the Galápagos Marine Reserve (GMR). Scientists also used photo identification and laser photogrammetry to monitor the hammerheads' movements and to document their daily behavior. The daily behavior of lions, another apex predator, has been well documented. Notorious nappers, they lounge during the day, sleeping as much as 18 hours, coming alive only at night to hunt. But do hammerheads behave like these land-based predators? What does their daily routine look like? Similar to lions, hammerheads spend their daytime hours passively. One of their favorite pastimes is resting. But scientists haven't yet determined whether hammerheads actually sleep like lions.

Hammerheads cannot stop moving because of the way they get their oxygen. The oldest species of sharks pumped water through their mouths and over their gills. This method is known as "buccal pumping," named for the buccal, or cheek, muscles that pull the water into the mouth and over the gills. Many sharks continue to practice buccal pumping today, including nurse and angel sharks, both of which lie on the bottom of the ocean floor. However, as sharks evolved, they found a better way to breath: ram ventilation. As a shark swims, it "rams" water into its mouth, letting it flow out through the slit of its gills. This method of breathing is more efficient, though it does require near-constant movement. Ram ventilators like hammerheads can't stop moving, or they die from a lack of oxygen. Since the hammerheads can't stop, it begs the question, can they sleep? Dolphins have to keep moving as well and still seem to sleep by shutting down one side of their brain while swimming. Perhaps hammerheads do the same; however, scientists have yet to confirm their theory.

The Galápagos study revealed that hammerheads live surprisingly like humans. We have daily routines, getting our morning coffee from the same place and commuting to work over familiar terrain. Like humans, hammerheads have their favorite places to hang out. The scientific term is "site fidelity," although a more colloquial term is "home base." Acoustic signals showed that the sharks linger in the same places around Wolf and Darwin Islands during the daylight hours. The hammerheads meander through the same reefs, gliding over the same yellow brain coral and maneuvering at low speeds along the colorful purple sea fans attached to the reef's steep slopes. As they hug the east side of these islands, they interact with other underwater denizens. Schools of red squirrelfish scamper into the reef's crevices and with big round eyes peer around back into the open water. The sharks also scatter the shoals of jackfish, who keep a wary eye on them. The sharks are not always lunging at prey, though they may come across the signal of a sick fish and surge forward to pick it off. That effort serves to stop the spread of disease on the reef, similar to preventing a sick human from boarding an airplane at the airport. But by and large, the hammerheads' day is spent passively on the reefs and slopes.

Then there is the night, when the underwater world transforms. A dramatic vertical migration takes place in the temperate regions of the world—a spectacle humans rarely see. Creatures like zooplankton rise up from the ocean depth under the starry skies. On the surface of the water, they feed on the phytoplankton, which convert the sun's energy into food. The tiny plankton that live among the waves dance on the surface and shimmer with bioluminescence.

In a single day, this vertical movement can range from 1,000 to more than 3,000 feet, depending on the size of the animal involved. Following the zooplankton are predators. At night, the biomass of the surface waters increases by as much as 30 percent. Squid and other fish prey on the zooplankton, and they all rise to the surface to feast, where the hammerheads greet them. Upon the dawn of a new day, the zooplankton retreat back to the depths for safety. As the zooplankton leave, the surviving squid beat a hasty retreat into the depths along with them. The hammerheads also return home after a night's work. A cycle that has been repeated for eons will play out again in the undersea theater the next night.

Another discovery is how hammerheads use the Galápagos Islands to separate adult from young hammerheads. Just as humans have set up areas for children to play in schools and for adults to work, the sharks have similar arrangements. Adult hammerheads live and work on the northern islands of the Galápagos. The southern islands are where the sharks have their birthing area. After a gestation period of ten to twelve months, females give birth to as many as twelve pups, although the great hammerhead can produce as many as forty. It is not known how soon the female hammerheads become pregnant again after giving birth. After the sharks are born, the researchers found, the baby sharks live in groups among the mangroves of the southern islands, feeding off the fish there. As they mature, they move north to join their adult peers, eventually venturing out into the open sea. On some of these recorded journeys, the sharks traveled 50 miles round-trip in a day. Moreover, the sharks traversed the entire way on a direct path. The scientists were puzzled about how they accomplished that feat. One theorized that hammerheads use the earth's magnetic field for navigation. When basalt hardens on seamounts, magnetic ridges and valleys form. By detecting the geomagnetic fields on the seafloor, the hammerheads can orient themselves. Whenever they need to reorient themselves, they take a deep open-water dive. One shark dove down to a depth of 3,000 feet and, using his head's electrosensory organs, was able to detect his position and head back home among the familiar rock outcroppings and reefs at his favorite home base on the island.

Occasionally, the hammerheads made longer trips from the north of the Galápagos Islands to other areas in the eastern Pacific. Some traveled more than 300 miles to the Cocos Islands. Other hammerheads were recorded making return migrations of more than 1,800 miles. Since hammerhead sharks are highly migratory and travel great distances, they are vulnerable on their travels to fishing fleets plowing the waves, which is why these migratory studies are important for determining areas that need protection, even in open waters, where the hammerheads are in transit.

The scalloped hammerhead, which is almost as large as the great hammerhead, developed mate selection through schooling, a highly unusual trait among other hammerhead species. Scalloped hammerheads can often be spotted swimming in large schools of a hundred or more. For many scuba divers, witnessing the splendor of schooling scalloped hammerheads is the holy grail. While this phenomenon takes place in the Galápagos and along the Great Barrier Reef, the best place to witness this event is around Costa Rica's Cocos Islands, which Jacques Cousteau called the best diving spot in the world. Around these islands, hammerheads swell into schools of well over a thousand, visiting the pinnacles like Dirty Rock, an outcrop of jagged rocks that drops down about 120 feet below the surface. Sea life bursts there. Among the staghorn and fire coral, divers cleave to the reef at 30 feet and stare out at a watery, deep-blue canvas, where sea life moves in and out to create an ever-changing seascape of brilliant colors and varying shapes of different fish and coral species. With their trailing stinger, spotted eagle rays cruise along the bottom of the ocean on graceful, undulating wings. Thousands of silvery jacks, whose scales flash in the sunlight, coalesce into a swirling globe so dense that divers completely disappear in them. Dolphins love to chase the jacks, but the fish are fast and keep eluding their pursuers. A hawksbill turtle with some green algae encrusted on her carapace skims the reef with her flippers. Hunting yellowfin tuna with their scimitar fins cruise in from the blue miasma. Sharks are everywhere on this reef. Half a dozen whitetip reef sharks nap on the seafloor. Galápagos sharks arrive in small groups and prospect for prey. Though these sharks live around the islands, they are found worldwide on reefs of oceanic islands, where they are usually the most abundant species of shark. Sleek and shaped like torpedoes, they can grow to 10 feet. They have come here for their appointment with the barberfish, a specialized fish that cleans migrating fish and many species of shark including hammerheads. These yellow fish are 8 inches long and shaped like a pancake with a black stripe running along the top. Their eyes are large, and their lips are tiny, puckered, and protruding, which allows the aptly named fish to nab parasites living on fish. A pencil-thin stripe between the eyes and lips completes their resemblance to a natty Italian barber. One hammerhead slows to a stop, and a dozen barberfish get to work on their client. They work the hammerhead's body and pick and stab at the crustaceans and other ectoparasites clinging to the shark. Eventually, the hammerhead moves on, and the next client comes into the cleaning station.

Hundreds of scalloped hammerheads slowly coalesce and sweep forward, their distinctive silhouettes tattooed against the rich background of endless blue. A living, breathing wall made of hammerheads creates a giant shoal of sharks. Peter A. Klimley, a behaviorist at the Bodega Marine Laboratory at UC Davis, has studied scalloped hammerheads for two decades at El Bajo, a seamount, or underwater mountain, in the Gulf of California. Klimley believes the hammerheads school for mate selection. In midday, after the sharks aggregate, a structural hierarchy forms according to size and age. The largest, fecund females force the smaller females out of the center of the school. Males fight their way to the center of the school to attract the prime females. Once the dominant females make their selection, the pairs separate from the group to mate.[8] The sexual encounters can be violent. The females get raked, and their skin can be in tatters after these encounters with the males. A year later, females give live birth to the next generation of hammerheads.

While this schooling practice is great for mating purposes, it also exposes scalloped hammerheads to death in some parts of the world. Costa Rica, in particular, is one of the world's most important participants in the shark-fin trade.[9] Finning is a common practice in the Cocos Islands, despite their designation as a national park. In 2012, the former president of Costa Rica, Laura Chinchilla Miranda, signed a law prohibiting shark-finning. But this law remains difficult to enforce because the Taiwanese mafia and, to a lesser extent, the Indonesian mafia maintain a strong—and intimidating—presence in the country, and they continue to export shark fins with impunity throughout the Pacific Rim. Mafia-controlled docks like Inversiones, Harezan, and others in the country's port of Puntarenas continue to export approximately 95 percent of Costa Rica's catches to Hong Kong.[10] The Chinese demand for shark-fin soup is insatiable. An estimated 1.7 million tons of sharks are captured worldwide annually just to keep up with this demand.[11] The sale of shark fins to the Chinese generates significant revenue for Costa Rica. The state's rich bounty of hammerhead sharks is worth significant dollars. Because of their high fin ray count, coupled with their towering dorsal fins and long pectoral fins, hammerheads are more highly valued than other shark species. Hammerhead sharks represent at least 4 percent of the fins auctioned in Hong Kong, the world's largest shark-fin trading center. Hauled onto commercial fishing boats, hammerheads are finned alive and then discarded overboard. Without a fin, the sharks struggle to stay afloat, eventually sinking to their death. Corpses litter the sea bottom.

It's not just finning, however, that puts sharks at risk. Various shark cartilage industries exist in Costa Rica. The country is a leading processer of raw cartilage, which they export around the world, most notably to the United States, for use by the medical supplement industry, which erroneously claims that shark cartilage can prevent cancer. Costa Rica processes up to 2.8 million sharks a year, with much of this carnage going to supply the shark cartilage companies. The idea that eating shark cartilage stops cancer started when *60 Minutes* aired a segment on the subject over twenty years ago. After the show, the myth sprang up that sharks do not get cancer, a ridiculous and scientifically unfounded claim.[12] Even as early as 1908, a captured blue shark was found upon dissection to have been riddled with cancerous tumors. In addition, eighteen other species of sharks have also been shown to have benign and malignant tumors. The medical literature is replete with studies that prove sharks do get cancer.[13] The University of Texas conducted a controlled, double-blind placebo study to determine whether shark cartilage is an effective treatment against cancer. All 379 patients in the study received radiation and chemotherapy in addition to a shark cartilage supplement or a placebo pill. The team found that those taking the supplement did not live any longer than those taking the placebo.[14] Previous studies, funded by the National Center for Complementary and Alternative Medicine, also found that shark cartilage did not benefit patients with lung or advanced breast or colon cancer.[15] And noted shark researcher David Shiffman in *Scientific American* remarked that eating shark cartilage is useless. "Sharks get cancer," he said. "Even if they didn't get cancer, eating shark products won't cure cancer any more than me eating Michael Jordan would make me better at basketball."[16] That same article shows a picture of a great white shark with a large tumor protruding from its lower jaw.

In summary, the evidence is clear: eating shark cartilage had no benefit at all. The disturbing myth that sharks don't get cancer has done great harm to both sharks and people.[17] The danger of this absurd belief is that cancer patients won't receive effective treatments. Moreover, as a cancer preventative, shark cartilage provides only a false sense of security. Other documented approaches[18]—such as getting screening tests, eating a plant-based diet, and exercising—actually lower the chances of getting cancer and may also help people with cancer control their disease. Sadly, the carnage continues. Sharks die unnecessarily for their cartilage, and with a $30 million industry at stake, some parties have every reason to see it continue.

Recreational fishing also poses a threat to hammerheads. More than five hundred species of sharks exist in the world, with some weighing a few pounds while others weigh a few thousand pounds. The largest ones are the most attractive to fishermen seeking a trophy. While hammerheads cannot compare in size and strength to tigers and great whites, they can weigh over 1,000 pounds. Given hammerheads' size and imposing dorsal fin, fishermen relish catching them, and charter companies like Mark the Shark's give them the opportunity to do so.

Misperceptions exist around catching sharks. One is that they are so strong and tough that if you release them by cutting the line, they will swim away and be just fine. In reality, sharks are at risk anytime they are caught since the struggle can lead to their death. They can swallow the hook or get banged on the side of the boat, which damages their organs. Scientific studies show that the struggle on the line many times leads to the shark's death, even if it is released. The University of Miami Abess Center for Ecosystem Science investigated the effects of catch-and-release fishing on shark mortality.[19] The study investigated experimentally simulated catch-and-release fishing on five shark species—hammerhead, blacktip, bull, lemon, and tiger sharks—in South Florida and Bahamian waters. Researchers took blood samples to examine stress, carbon dioxide, and lactate levels. Researchers then used satellite tags to determine postrelease survival. Not surprisingly, the blood lactate levels of sharks soared after fighting on a fishing line in much the same way lactate in humans rises during intense or exhaustive physical exercise. An increase in lactate levels significantly affects the odds of survivability of many fish species. The study showed that even with minimal fighting times on a line, hammerheads exhibited the highest buildup of lactic acid. The lead author, Austin Gallagher, wrote, "Our results show that while some species, like tiger sharks, can sustain and even recover from minimal catch and release fishing, other sharks, such as hammerheads, are more sensitive."

With the ubiquity of GoPro cameras, YouTube abounds with videos and pictures of fishermen catching hammerheads. One YouTube video clearly shows the vulnerability of the hammerhead to sports fishing.[20] One hammerhead, filmed underwater, was caught on the line and was pulled up next to the boat. It was not long before a tiger shark appeared on the scene due to the hammerhead's electrical distress signals. In the fight between the tiger and the hammerhead, the fishing line holding the hammerhead snapped. However, the hammerhead was so exhausted that it could do nothing to stop the tiger shark from continuing its attack. The tiger shark took a bite out of the hammerhead, and the water turned crimson with blood. The tiger shark attacked again at the hammerhead's midsection and dragged the hammerhead into the depths beyond the camera's view.

In another case, a fisherman hooked a 14-foot hammerhead shark near Corpus Christi, Texas, a feat he called a "catch of multiple lifetimes."[21] The fisherman claims to have tried to release the shark back into the ocean, but it was "too tired" and did not survive. Of course, the death of the hammerhead comes as no surprise after looking at the data from the University of Miami. While the fisherman tried to show remorse that the shark did not survive, many on Facebook expressed horror at the tragedy. As one commenter on Poco Cedillo's Facebook page noted, "Had they not been sport fishing, that shark wouldn't have died that day."

In the United States, both the commercial coastal fisheries and the pelagic longline fishing industry punish the hammerhead population.[22] Recreational fishing is also having an impact on the shark. Hammerheads are the third most common shark caught, according to reports from Florida recreational charter companies, and their clients consider the great hammerheads specifically to be one of the most attractive species to catch. Across the coastal states of the Carolinas, Virginia, and Florida, hammerhead populations have been decimated. But the United States is not alone in its persecution of hammerheads. The global population of the great hammerhead shark is estimated to have declined by approximately 80 percent over the past twenty-five years. Scalloped hammerhead sharks have declined by 89 percent over a fifteen-year time period, from 1986 to 2000, less than one generation.[23] As a result, since 2007, the great hammerhead has been listed as endangered by the International Union for Conservation of Nature. The shark has also recently been included in the Convention on International Trade in Endangered Species Appendix II, which includes species not necessarily threatened with extinction, but in which trade must be controlled in order to maintain their survival. Yet in spite of the human onslaught on sharks, hammerheads rarely attack humans. They are considered neither aggressive nor dangerous to humans. Only seventeen unprovoked attacks by hammerhead sharks on humans have been documented. No human fatalities have been recorded over the past four centuries.[24]

Hammerheads are also faced with new challenges arising from global warming. The world's seas are approximately 1.3°F (1°C) warmer than they were a century ago, mainly due to the increase in CO_2. This fact is changing the ecosystem within which the sharks live in unknown ways. For instance, blacktip sharks usually migrate from the Carolinas into South Florida's waters as part of their annual winter migration. The sharks normally spend between mid-January and mid-March off the South Florida coast in search of warmer water and food. As the water has warmed, however, the blacktips have no incentive to move into the waters off the southern tip of Florida. What this means to the hammerheads, who hunt the blacktips, and the ecosystem of South Florida is unknown.

With such a perilous present and future, the hammerhead remains vulnerable, which puts into jeopardy the opportunity to learn from the species's many remaining mysteries. Does the hammerhead have more ampullae than other sharks, given its wider head? Is the head teeming with senses that are superior to those of other sharks? Could the tubercles assist with the shark's yaw (sideways turning motion)? New questions surface all the time. For instance, bonnethead sharks have been found to feed on seagrass, which sometimes makes up as much as half their stomach contents. Even though they are able to partially digest the grass, do they swallow it intentionally? That preference would make the hammerhead an omnivore—the only known omnivore among shark species. Scientists will be exploring the answers to these questions for many years.

But here is what we do know. Hammerhead sharks migrate extensively, and they face constant risks from humans in oceans around the world. Commercial fishermen have devised all kinds of nets. One of the deadliest is the drift net, which is left to drift in coastal waters where various fish species get entangled. When a hammerhead struggles to free itself, the shark only gets more entangled and, once immobilized, will suffocate to death. Since hammerheads travel mostly in coastal waters, they often fall victim to these nets. Moreover, recreational anglers pursue the hammerhead as a trophy. In the United States, some of the 49 million licensed recreational anglers, armed with equipment worth $45 billion,[25] descend on sharks, many of them hammerheads.

While hammerheads are hunted constantly, their lives are quite ordinary; they live in their home territories, find mates, reproduce, and die. In that sense, there is little difference between the life of the hammerhead and the life of humans. Yet humans are different, and we have unique attributes. Darwin made the distinction clear. He said, "The love for all living creatures is the most noble attribute of man." Humankind needs to widen its circle of love to encompass sharks now more than ever. However, while humankind can be noble, we can also be victims of self-delusion. As Darwin also observed: "Great is the power of steady misrepresentation." The misrepresentation of sharks has led to the greatest threat to the hammerhead's existence in 20 million years. And as we'll see in the next chapter, our misconceptions about sharks as solitary, bloodthirsty species continue to be upended. Scientists continue to discover that sharks have more in common with people than previously believed.

Chapter 4
Sharks As Social Animals

At Shark-Con, an annual event held in Tampa Bay, I met a friendly man named Duncan Brake, who travels the world filming sharks and other marine wildlife. His work has taken him everywhere, from the Falklands to Antarctica. He told me excitedly about a project he was working on in the mangroves of the Bahamas, where scientists are finding out all sorts of things about the social behavior of lemon sharks. I asked Brake to put me in touch with these scientists, so he introduced me to Tristan Guttridge, executive director of the Bimini Biological Field Station, otherwise known as Sharklab. The lab houses a number of scientists conducting experiments and exploring shark behavior. Populated with various shark species, Bimini is an ideal place to conduct shark research because it's located 45 miles east of Fort Lauderdale, and relatively untouched—far from the interests of tourists or developers.

Young and energetic, Guttridge strikes an interesting picture: with his wavy brown hair, impish smile, and two-day growth on his handsome face, he projects a vibe that's more beach bum than shark expert. During our conversations, though, his seriousness and passion for sharks revealed the accomplished scientist and postdoc lurking under his casual exterior.

As far back as Guttridge can remember, he has been mesmerized by sharks. His grandfather, a Navy veteran of World War II, introduced him to the world of nature, regaling him with stories. In one of them, he described his run-in with a shark after his aircraft was shot down off the coast of Malaysia. Guttridge's grandfather and another survivor exited the plane before it sank, and they made a mad dash for the beach. "I saw a dorsal fin as big as a block of flats," his grandfather said. The shark circled them and, despite showing great curiosity in the two airmen, allowed them to reach the beach without incident. The story stuck with Guttridge, who cultivated his love of nature.

Like a young Indiana Jones, Guttridge traveled, at the age of eighteen, to Madagascar, where he conducted coral reef surveys, working amid a school of barracudas.

Once he finished the six-month survey, he moved on to Tanzania to participate in a terrestrial biodiversity project. While in Tanzania, he was stung by a scorpion and, after accidentally stepping on an ant trail, had his legs and genitals overrun by them. He also climbed Mount Kilimanjaro. Following these adventures around the world, he returned home to the United States to volunteer at Sharklab, investigating the social organization and behavior of lemon sharks, which congregate in mangroves and shallow subtropical waters. In 2010, he earned his PhD at the University of Leeds. There Guttridge worked with another scientist, Culum Brown, a professor of biological sciences at Macquarie University in Australia, coeditor of the book *Fish Cognition and Behavior*, and editor of the journal *Animal Behavior*. With Brown, Guttridge uncovered new information about the traveling patterns of lemon sharks.

In a groundbreaking study, Guttridge proved that sharks—in this case, lemon sharks—are not the solitary animals that popular opinion assumes they are. None of the shark books I scooped up as a child, for instance, ever touched on the idea that sharks exhibit social behavior, only instincts. As it turns out, though, they are social animals, just like humans. Eager to prove this point, Guttridge set up an experiment with sharks in long pens separated by a plastic mesh. The sharks could see each other, but they couldn't come in contact with one another. Guttridge and his team installed a single shark in the middle pen (zone 2) and two to four sharks in the far pen (zone 3). They kept the nearest pen (zone 1) empty. In each test, the solitary shark in zone 2 preferred to spend time near zone 3, where its fellow sharks congregated, demonstrating an active preference to socialize. Without food or any other potential motivator, it was clear that the sharks were seeking a connection with their fellow sharks. Guttridge was excited. He realized he had come upon a new area in shark research, every biological scientist's dream. His study proved that sharks clearly demonstrate sociality and cognition, something heretofore never considered in the species. "When given the choice," Guttridge said, "sharks had an active preference to be social."

Guttridge did more than design and execute his study. He started spending time in the mangroves to observe lemon sharks up close as a kind of subtropical Jane Goodall. Mangroves are a key habitat for lemons because they offer juvenile sharks cover while they develop hunting and other survival skills. In the open ocean, juvenile sharks are defenseless against larger predators. Mangroves are also important ecosystems for sea life. Mangrove trees, which have thick, multiple roots anchored in the mud, provide a hiding place for a cornucopia of animals. The thicket of branches, densely packed with small green leaves above the roots, can extend 100 feet in the air. These branches are rookeries, or nesting areas, for coastal birds like herons and egrets. Kingfishers, with their translucent blue plumage, camp along the branches of the black mangroves, where they wait to spear small fish with their sharp beaks.

Guttridge set up a watchtower under the blazing sun in the Bahamian mangroves. In this Eden, far from the intrusions of humans, he studied the sharks, surrounded only by natural wildlife. From his tower, he watched the sharks and other indigenous sea creatures swimming and twirling around the mangrove trees' horizontal brown roots. While music and news from his transistor radio played in the background, he took copious notes, sometimes watching for days at a time. Eventually, the days turned into months, which somehow added up to three years in the mangroves. This time was well spent, because Guttridge discovered that the sharks were doing more than simply aggregating in the mangroves. They were establishing social bonds with one another that extended beyond hunting assistance and mate selection. He observed sharks spending time with their compatriots, sharks of similar size and age, like boys and girls do in school. In addition,

Guttridge witnessed juvenile lemon sharks regularly traveling together through the mangroves in a sort of carpool, tucked closely together as a single unit.

What struck Guttridge most was how lemon sharks behaved during high tide, a dangerous time in the mangroves. Because the incoming tide raises the water level, predators can move freely into the fringes of the mangroves, seizing vulnerable prey like juvenile lemon sharks. The gimlet-eyed barracuda, for example, which is pocked with black, is known to steal into the shallows to nab unsuspecting lemon sharks. Remarkably, Guttridge observed lemon sharks seeking shelter from barracudas and other opportunistic predators in pairs, partners in survival. Once the juveniles knew they were safe—when high tide was over and the water level drifted back down—the sharks would disperse.

When juvenile lemon sharks weren't working together, they played together, which allowed Guttridge and his team to note that individual sharks exhibit distinct and diverse personalities. Some sharks, according to Guttridge, like to explore and cover broad territories, while other sharks prefer to stay closer to select sites. To document this behavior, he teamed up with Duncan Brake, who set about capturing it on film. Because the underwater roots and the tall branches of the mangrove trees blocked his desired shot, Brake sent a drone over the top of the forest. He was astonished at what he saw. Wide and deep canals thread their way through the mangroves, serving as a kind of underwater highway for the sharks. The sand at the bottom of the canals and the shallow water allowed the scientists to see the sharks clearly in the light-green water and film them as they traveled freely through the canals. In a typical video, two lemon sharks, each about 3 feet long, swim as a pair. Sometimes they meander along, and sometimes they speed up and, with a few flicks of their tails, dive together in and around the mangrove roots. Sometimes one ventures across an underwater branch, its cartilaginous friend in hot pursuit. The scientists sometimes saw half a dozen lemon sharks circling each other in congregations. Because Brake didn't observe the lemon sharks hunting or feeding, he was sure their behavior was strictly social, a theory Guttridge backed up.

The lemon sharks' ability to socialize prompts an obvious question: If they interact so freely, can they learn from one another? Since these young sharks need to learn how to hunt, for example, is it possible that other lemon sharks could teach them? How animals learn to hunt has been extensively studied in other species, but no such studies have been done on sharks, until now.

To investigate the social learning capabilities of juvenile lemon sharks, Guttridge designed another experiment, creating a novel, but simple, food task. Sharks entered a start zone, where they came in contact with a target that dispensed a food reward. Sharks were trained to get the reward, and then they acted as "demonstrators" who knew how to get the food. Guttridge paired these sharks with untrained juveniles. To control the experiment, he also allowed untrained juveniles to interact with "sham demonstrators," sharks that had no previous experience with the task. Guttridge recorded both groups and then compared the results. The sharks working with the "demonstrators" learned the correct steps to get the reward, while the sharks paired with "sham demonstrators" failed. By observing other sharks in their group, the juvenile lemon sharks learned new tasks, which proved to Guttridge and his team that juvenile lemon sharks can apply socially derived information to their environment. Guttridge's mentor and coauthor, Culum Brown, summed up the research's conclusion: "I personally work on spatial learning in fishes quite a bit, and I think it is fair to say that sharks' abilities in this area are just as good as any vertebrate."

Brown and Guttridge also discovered that lemon sharks in Bimini have a strong site fidelity, or preference, for this island. If one were to encounter a lemon shark 5 feet in length or seven years of age, there is a 50 percent chance it was born in Bimini.

What Guttridge recorded has changed how science looks at sharks. Darwin said that humans and other animals share the same attributes, the only difference being the degree of those attributes. Humans are clearly the most sophisticated species in terms of creating connections with others. However, this feature does not take away from the connections formed by other species, like sharks. We must now view sharks in a new way: as creatures that share our attributes, specifically intelligence and sociability.

Because Guttridge worked with Culum Brown on the study, I wanted to learn about Brown's shark research in Australia, on the other side of the globe. I called his university office in Sydney. Our conversation focused on Port Jackson sharks, a bottom-dwelling species found only in Australia. Port Jacksons have a high forehead and a gray camouflage color with black lines scattered across their top. They have strong jaws and back teeth that can crush hard-shelled animals like mussels and sea urchins. Stocky and built like miniature tanks, they grow to about 5 feet in length and are, by far, the most common shark in Australian waters. These qualities, in addition to their relative passivity, make Port Jacksons an easy species to tag and release.

Brown and his team captured sharks and attached

tags to their dorsal fins. They then made a small incision on the underside of each shark and inserted an acoustic tag designed to send out a ping whenever the Port Jackson shark came within a third of a mile of an underwater receiver or within 30 feet of another tagged shark. Each ping was time-stamped. Using these proximity-tagging methods, Brown was able to track the sharks on a map, and the scientists could identify each shark individually and, therefore, monitor interactions among the sharks. In other words, Brown and his team could determine which sharks hung out together.

Brown released the tagged sharks in Jervis Bay on the New South Wales coast, 90 miles south of Sydney, an ideal spot to study their behavior because it hosts several seasonal Port Jackson mating aggregations. He followed the migrations of those small sharks south from Jervis Bay to Tasmania, a trip of approximately 600 miles. Several findings came out of the study. Some sharks returned from their 1,200-mile round-trip journey each winter for five consecutive years, and they returned to the same location—not just the same bay or body of water but also the same reef from which they started. These migrations showed the incredible accuracy of the shark's homing ability.

Brown also observed sharks spending most of their total aggregation time with sharks in their own social network. Like the lemon sharks in Bimini, the Port Jackson sharks liked to stay with sharks of the same size and sex. Some were recorded staying with the same sharks over a couple of years. "Large groups of sharks formed in the breeding season is not just a random collection of individuals," Brown said. "The sharks preferred to hang out with other individuals who were similar to them." Brown's coauthor, Jo Day, a researcher at the Taronga Zoo, said, "The sharks established long-term relationships over many years." One researcher compared the paper's findings to Marlin's and Dory's behavior in *Finding Dory*.

Later, Brown told me that the intelligence of sharks is underrated. Their navigation skills are superb and they learn survival skills from one another. If they have all these remarkable senses combined with intelligence, then the next logical step is to conclude that sharks—and other fish—feel pain. A human's physical ability to feel pain is a trait inherited from a fishlike ancestor. The nerves endings and related structures are identical. The perception of pain, and the associated psychological response, evolved to protect animals from harm. "It would be impossible for fish to survive as the cognitively and behaviorally complex animals they are without a capacity to feel pain," Brown told *HuffPost*.[1]

If Brown is correct, sharks suffer a great deal of pain

when they're caught, gaffed, and finned. Death does not come easy to sharks. When they are finned alive, they slowly suffocate to death, which can take anywhere from ten to twenty minutes. Brown believes that a review of the evidence for pain perception strongly suggests that fish experience pain in a manner similar to the rest of the vertebrates. Although no one can provide a definitive answer on the level of fish or shark consciousness, the extensive evidence of fish cognitive sophistication and pain perception suggests that the best practice would be to lend fish like sharks some level of protection against cruelty, as society would to any other vertebrate.[2]

The curtain is slowly being drawn back to reveal an unappreciated aspect of sharks. The study of sharks is just now beginning to uncover real breakthroughs. It wasn't until the turn of the twenty-first century that scientists realized that several species of sharks aggregated for mating and predation-avoidance purposes. A 2004 aerial study proved to researchers that basking sharks aggregate,[3] while a 2013 study using active and passive telemetry showed leopard sharks engaged in the same aggregating behavior.[4]

In the years since, scientists like Guttridge and Brown have expanded our knowledge of sharks, including the misunderstood species' long-overlooked ability to engage in social behavior and establish lasting, meaningful relationships with one another. With a high brain-to-body-mass ratio compared to that of other fish, sharks have the intelligence to engage in complex social behaviors, such as forming dominance hierarchies and creating social bonds.[5] Of course, different shark species vary in social abilities, but even "solitary" sharks like the great white and the hammerhead might in fact integrate some aspect of social interaction into their behavior.

The sociability of all animals lies on a continuum from highly social on one end to completely solitary on the other. And on this continuum, the common perception of sharks has held that they are strictly solitary creatures. Of course, where sharks land on the spectrum depends on the species. Because great whites and tigers are strong and savvy enough to survive on their own, they remain fiercely independent hunters. Many smaller shark species, on the other hand, are social creatures. At least two species, lemon and Port Jackson sharks, enjoy the benefits of being in social groups.

As I traveled on my journey, I came across evidence in other shark species that supports Guttridge's conclusions about the social behavior of lemon sharks. Whale sharks, for example, have been seen aggregating in the

Gulf of Mexico during the summer months, as well as off the coast of Kenya. Scientists have recorded basking sharks, which are large filter feeder sharks, near Newfoundland and swimming in large groups of more than a thousand individuals. The extent of the social interactions during these aggregations has not been fully studied, but is it possible that some socialization is taking place?

Some research also points to blue sharks hunting together in small packs. Researchers have observed blue sharks working together to feast on schools of anchovies. One group of blues drives the anchovies into "bait balls," and then the other blues take turns lunging at the fish, although more studies are necessary to draw definitive conclusions.

All creatures have to learn from birth how to hunt in a challenging world. Science has revealed the dynamism of the shark species in that there is no "one size fits all" behavior. Sharks' great strength is their flexibility to alter behavior among the species to survive. I learned along my journey that great whites are autodidacts, learning on their own how to hunt fish and later seals. In South Africa, great whites soar out of the ocean depths to catch seals, demonstrating that, by the time they mature, they master their self-taught lessons. Now I have discovered that some smaller shark species developed another approach to learning how to hunt, the ultimate life-survival skill. Through evolution, smaller sharks developed an adaptation to learn from one another. Unlike any other species, the shark's remarkable adaptability has allowed them to flourish.

This relatively new discovery of the shark's ability to communicate requires a reassessment of shark intelligence. The fact that some sharks have the ability to learn from other sharks shows that they must have the intelligence to communicate and absorb complex information. This intelligence to connect with their fellow creatures should not come as a surprise: apex predators have to be intelligent to survive. When the lemon sharks leave the mangroves as adults, they can literally thank their peers. Each new generation of lemon sharks moving out from the mangroves into the open ocean is a tribute to a dynamic and malleable species that has withstood countless upheavals over 450 million years to rule the seas.

Chapter 5
The Quest for
the Tiger Shark

My quest to dive with a tiger shark took me to the Hawaiian Islands, where tigers are abundant. Scientists like Carl Meyer, PhD, of the University of Hawaii have been studying tigers for the past twenty years, uncovering exciting new information about one of the few shark species big and strong enough to play a major role in protecting the marine ecosystems around the world.

I traveled to Honolulu on the island of Oahu in Hawaii. The most southern part of the United States, Hawaii has one of the longest coastlines in the country; only Alaska, Florida, and California enjoy longer coastlines. Everywhere I looked, luxurious green mountains soared over a blue sea. The tallest mountain, Mauna Kea—on the Big Island—stands 13,800 feet above sea level. However, if you measure the base from the seafloor, it rises 33,500 feet, which makes it taller than Mount Everest.

Meyer asked me where I wanted to meet, and I suggested the North Shore solely because that's where world-class surfing takes place. Storms in the Pacific Ocean create winds that whip up the waves. As the waves barrel over thousands of miles, they begin to merge and grow in strength. With no impediment to their movement, they can reach heights of 40 to 50 feet in winter, equal to the length of a semitrailer truck. So I headed to the North Shore and found seven miles of pristine coastline. The waves were calm.

I was immediately struck by the remarkable way Meyer articulates the issues facing sharks today. "The broad goal of our research," he explained, "is to learn more about the natural ecology of tiger sharks, and other species of sharks, that are found in Hawaii. It's important to understand the ecology of these animals in order to develop effective conservation strategies for these important top predators. We use electronic devices to determine where these animals travel in the ocean, and the types of habitat that they use. We're also interested in learning things like how fast they grow, how often they eat, and where they reproduce."

Tiger sharks prefer to live in tropical and moderate coastal regions, thriving in murky waters and estuaries. The tiger shark's name does not come from the aggressiveness of the shark but from the pattern of dark stripes along the sides of its body. The tigerlike stripes are most prominent during the shark's youth, when it is vulnerable. Because tiger sharks stay close to the coast, the camouflage turns them into meandering shadows directly below the water's surface. When the sharks reach adulthood, the stripes slowly disappear. The tiger is one of a kind. Its broad head, which is almost square when seen from above, gives the shark the chiseled-jaw look of a movie star. Its nostrils, called "nares," are quite pronounced,[1] splayed in front of the snout. Tucked inside the tiger's eye is the nictitating membrane, which covers the eye for protection when the shark is feeding. This same membrane is found in many shark species, but there is an important difference between the eyes of tigers and those of other sharks. The tiger's sclera, or white of the eye, encloses its black center. While the mako's eye is completely black, which gives the shark a serious, "I mean business" look, the more soulful tiger's eye is closer in its color composition to a human eye (a white circle with a black center), which gives the shark a more thoughtful, sometimes mischievous, demeanor. This is appropriate because the tiger is an ambusher and a thinker, absorb-

ing its surroundings. Tiger sharks have made many ap-
pearances in YouTube videos in which scuba divers are
petting their noses and the sharks are seemingly enjoying it. The eyes in these video clips have an impish look
with a hint of playfulness. Natives of the Hawaiian Islands call the tiger shark "*niuhi*" and hold the shark's
eyes in reverence. Legend has it that many Hawaiian
kings ate *niuhi* eyes to help them predict the future.
The mother of King Kamehameha, the most famous
Hawaiian king, requested *niuhi* eyes during her pregnancy because she believed they would help the future
king become a better leader.

Despite the tiger shark's folkloric ability to foretell the
future, when it comes to hunting, the tiger shark remains
the emperor of the ambush. The Big Four sharks—
the great white, tiger, mako, and hammerhead—all take
a different approach to hunting. The attack sharks, the
great white and the mako, capitalize on their blazing
speed to catch underwater prey and use their powerful
jaws to settle the matter with a clear-cut finality. The
tiger is different. Its Latin name, *Galeocerdo cuvier*,
hints at the shark's true nature: cunning. (*Galeocerdo*
translates roughly as "fox shark," and *cuvier* refers to
the surname of eighteenth-century French natural-
ist Jean Léopold Nicolas Frédéric, Baron Cuvier, also
known as Georges Cuvier, who came up with the idea of

classifying animals into different phyla, or kingdoms.) Capable of swift and deadly attacks, tiger sharks possess the speed to catch a wily dolphin. For the most part they swim slowly and methodically, waiting and sensing for the right opportunity to strike since they are at their core an ambush predator.

Innately curious, the tiger is known to come right up to a diver's cage, even to the divers themselves, because underwater cameras generate an easily detectable electrical signal, which draws in the nosy shark. Tiger sharks gather information with their mouth, which is why divers regularly misinterpret a bite to their camera as an act of aggression, rather than the harmless, exploratory gesture it is. Tiger sharks like to sneak up on divers, disappearing and reappearing like a magician's trick, which unnerves many.

Another way to differentiate tigers from other shark species is their teeth. Like most sharks, a tiger replaces its rows of new teeth repeatedly throughout its life. But that is where the similarities in dentition end. While other shark teeth grab and hold prey, like kitchen forks, the tiger's teeth act more like diamond cutters and are composed of two parts. The forward part of the tooth is shaped like an A, and its razor-sharp tip is serrated for cutting into its prey. The back part of the tooth is flat, with a slight downward curve. These serrated teeth can crush bone, saw muscle, and rip fins. The tiger's teeth are a weapon of mass fish destruction. According to numerous tests, the tiger's jaw can generate a force of 3 tons per square centimeter, which is equal to the weight of two cars. For almost all creatures, a turtle shell is practically impossible to penetrate. But a tiger shark can smash the tough carapace, turning the turtle into a heap of blood and sinew. The shark's bristling arsenal of twenty-four identical teeth in its upper and lower jaws is an evolutionary triumph. And tigers need those teeth because they have the broadest diet of any shark species.

"They eat all kinds of reef fishes. They eat many types of invertebrates—things like lobsters and big marine snails," Meyer told me. "They eat a lot of octopus, they eat birds, they eat marine mammals, they eat other sharks, they eat turtles, they eat sea snakes. They eat all kinds of stuff that washes out into the ocean—dead cats, dead cows, bags of skinless, boneless chicken. So, that really sets them apart from a lot of other sharks, which have much narrower diets."

Sometimes, however, what tigers eat is rather shocking.

In 1935, in New South Wales, Australia, commercial fishermen accidentally caught a 1,600-pound tiger shark. Because they weren't hunting the shark, they set it free in a tank, while they arranged delivery to the Sydney Aquarium, which requested the shark for an exhibit. Sharks do not do well in captivity, and this tiger shark refused to eat for a week. Eventually, he vomited up several objects, one of which was a human arm. The appendage featured a tattoo of two boxers on the shoulder. The police conducted an investigation and, based on the tattoo, identified the victim. Someone had killed the man, chopped him up with a knife, and thrown the severed body parts into Sydney Harbour, where the tiger shark had found them. The police investigation ended there because, according to Australian law, a single arm did not constitute a murder. The shark didn't get off as easily, however, because the aquarium decided he was too difficult to keep. The director ordered his staff to butcher the shark, a second innocent victim in this bizarre tale of murder.

Since this curious case, researchers and fishermen have continued to find a number of other strange items in tiger stomachs: boat cushions, tin cans, license plates, tires, and the head of a crocodile. The shark is a high-tech machine assigned the modest job of ocean cleanup. The ocean owes a debt of gratitude to the tiger shark for maintaining it. When tigers remove garbage—weak and sick fish—they remove from the ocean bacteria and viruses that can harm reefs and seagrass. However, the tiger's work extends beyond mere custodial work: as apex predators, tiger sharks play an important role in maintaining the balance of fish species across the ecosystem. Moreover, the research shows that areas with more apex predators have greater biodiversity and higher densities of individuals than do areas with fewer apex predators.[2]

Tiger sharks are fast growers. On average, tigers reach 12 feet by age seven. The fastest-growing tigers are this long by age four. In human terms, this is like a toddler being the size of a sumo wrestler. Other species of sharks grow only a few inches per year. In addition, tigers are among the largest of the various shark species. The larger tigers usually grow to a length of 18 feet, about the height of a giraffe, but some may reach more than 22 feet.

Tigers and humans occasionally interact, most commonly when skin divers harpoon fish. The fish's telltale electrical distress signal attracts tigers. The sharks are more interested in getting a free lunch than they are in the divers, and shark attacks on humans are rare.

They still do occur, however. According to Hawaii's Department of Land and Natural Resources (DLNR), there were nineteen confirmed shark incidents reported around the islands of Oahu and Hawaii, between 1995 and 2015, which is less than one per year. Around Maui, there were thirty-eight confirmed incidents, or less than two per year. Though Oahu and Hawaii are six times more populous than Maui, the less densely populated island experienced twice as many shark attacks. In 2012 and 2013, twenty-three bite incidents occurred, a threefold increase above the average rate of attacks. Again, most of these attacks occurred in the waters around Maui.

The DLNR's Division of Aquatic Resources recruited Carl Meyer to help them figure out why more attacks were occurring in Maui and what, if anything, Hawaiian state authorities could do to protect beachgoers and vacationers there. Highly regarded in Hawaii and on the mainland for his innovative research on the ecology and management of sharks and reef fish, Meyer studies movement patterns, habitat use, and trophic ecology of sharks and fish, as well as the navigational abilities of sharks. To more closely monitor the sharks around all three islands, he captured and fitted twenty-four tigers with pop-up archival transmitting tags (PATs), which can record data for many weeks or months without transmitting. At a preprogrammed time, the tag pops off, floats to the surface, and starts an immediate data dump, transmitting summary information to satellites. While the summary data is still valuable, it is incomplete. Unless scientists physically retrieve the tag, they can't access the full data set each PAT contains. And if the battery dies before the tag is retrieved, data will be lost.

Meyer waited for the tags to pop up. When one PAT rose to the surface, its signal went out to a lab at the University of Hawaii. Right away, lab technicians notified Meyer, who had already arranged for a boat team to go out and retrieve the tag. There was a problem, however: severe weather was preventing the team's small 15-foot boat from venturing out to sea. The landlocked scientists circulated the tag's geolocation to fishermen in the area. None of them spotted the tag, and the clock was ticking. Fortunately, the weather improved enough to risk letting the Hawaiian team motor out 43 nautical miles east in the middle of the rough seas. They used a special directional scanner to locate the signal, but the 4- to 6-foot waves made it difficult to spot the small tag. With only two minutes remaining before the captain's predetermined turnaround time, the team spotted the tag off the coast of Maui and brought it aboard. The mission was accomplished, and none too soon—the satellite tag battery had only twelve hours of power left. If they had waited one more day, the tag's data would have been lost forever.

The information from that tag, combined with other

material the team collected, allowed Meyer to piece to-gether an extraordinary picture of the tiger shark's trav-els and revealed, per the DLNR's mandate, the reason why shark attacks occurred more frequently in Maui. Tiger sharks prefer certain areas over others—one of which is the coastal shelf, the relatively shallow water surrounding the Hawaiian Islands where the water is less than 600 feet deep. This area is more conducive for feeding and reproducing, according to Meyer. And because there is considerably more of this coastal shelf habitat around Maui than any other Hawaiian island, more tiger sharks gathered there. Meyer noted that the tigers he tagged in the waters off Oahu traveled to the coastal shelf off Maui, while the Maui tigers he tagged stayed put. A significant gathering of tigers formed there, which is why Maui experienced more attacks.

Increase the number of sharks in a specific area and the likelihood of an attack increases. That's just math. Still, the actual rate of tiger shark bites is incredibly low when considering that tigers regularly approach the beach, often in only a couple of feet of water. "It just so happens that the areas that are the most attrac-tive to tiger sharks are also the places where humans like to play in the ocean," said Meyer, who described the spike in attacks between 2012 and 2013 as a statisti-cal anomaly, a one-year aberration. "From a strictly statistical perspective," he said, "twenty-three incidents in two years can just be chalked up to 'the elements of chance'—especially considering how many people in Hawaii partake in ocean activities—as opposed to some environmental trigger or some other trigger."

Rather than closing the beaches of Maui—or, worse, launching an ill-advised shark-culling campaign—he instead recommended that Hawaiian state officials keep the beaches open. Sure enough, the number of attacks returned to the mean. In 2018, there were only three confirmed shark bite incidents in the state.

"In spite of the shark's size, humans do not need to fear them," he said. "Shark bite incidents are very, very rare, and are actually very unusual behavior. If tiger sharks really wanted humans as prey, it just wouldn't be safe to go swimming."

Since the state ended its study of shark attacks in 2016, Meyer has continued to learn a tremendous amount about the travels of the tiger sharks from the tagging studies. "Tiger sharks, in particular, are very cosmopolitan in their movements," he said. "They range everywhere from the shallow waters right next to beaches, to far out into open ocean—sometimes, more than 1,000 miles out into the open Pacific Ocean. And they also range between the surface and depths of around 1,000 meters—that's around 3,000 feet. And because they're ranging between the surface and these depths of 3,000 feet, they're encountering temperatures that range from tropical warm oceans down to very cold temperatures that are similar to the Arctic Circle."

Meyer uses the same kinds of cutting-edge technologies to study other coastal sharks in Hawaii, including the sandbar shark, a common species in the region. These sharks are relatively small; a large adult is only about 6 feet long. Meyer captured a sandbar shark and ingeniously fitted it with a video camera to find out what it was up to. When Meyer recovered the camera, he saw that the shark swam out a couple of miles offshore and, after a couple of days, went down to a depth of about 300 feet and joined a huge aggregation of sharks. "This aggregation consisted of a variety of species, including other sandbar sharks, blacktip sharks, but also large numbers of scalloped hammerhead sharks," Meyer said. The group of sharks gradually moved from 300 feet in depth up toward the surface; at dusk, the aggregation dispersed, and the sharks went their separate ways. "The most likely theory is that this sort of phenomenon is actually an antipredation mechanism. So although we're talking about sharks, who are relatively large—anywhere from 4 or 5 feet up to, maybe, 10 feet in length—they are still subject to predation by larger sharks, like tiger sharks. About 40 percent of the diet of the largest tiger sharks is made up of other sharks. So, we think that these mixed-species aggregations form for the purpose of protecting the sharks within the aggregation from predation from large species."

Meyer is convinced that fear is the driving force in these aggregations. If this was a mating aggregation, which is another reason why sharks aggregate, then the scientists would expect to see only one species of shark in the pods because one species, like sandbar sharks, cannot mate with another, like scalloped hammerheads. I used to think that sharks were the one animal above having to fear living in the ocean, but fear is ubiquitous in the underwater world. The sharks revealed an ability to come up with a strategy for survival with their compatriots. Their behavior shows a consciousness of themselves and others and an ability to collectively figure out a way to survive in a harsh environment.

When it comes to sex and reproduction, scientists are still learning about the breeding habits of tiger sharks. One thing they do know is that tiger sharks have a large number of pups, more than most other shark species. Tiger sharks are ovoviviparous, which means they have eggs that hatch inside the womb so they give birth to live young, usually about thirty or forty pups at a time, though tigers sometimes birth as many as eighty pups. But with a gestation period of fifteen to eighteen months,[3] tigers take a long time to reproduce. While they are protected from overfishing in Hawaii, the sharks are in peril in many other parts of the world.

Because I wanted to get close to a tiger shark, I signed up for a cage-diving trip in Maui and soon was on the deck of a 45-foot powerboat. I was not alone in my fascination with sharks. A dozen other people joined me for the dive. We were all ready to dangle in a cage floating just below the surface to get a glimpse of sharks. Each person that I talked to had their own reason for participating in the adventure. Some were daredevils, while others were more like me: they wanted to see the beauty of sharks up close. The day was sparklingly clear, and as we headed out to sea, the green mountains of Maui faded. I am used to the harsher look of the Atlantic Ocean. Its green waters are always seemingly dressed with white caps. In contrast, the Pacific that day was unusually calm, its blue color vibrant and warm. When one is on the Pacific, the ocean whispers to the soul.

I talked with one of the guides about her experience diving with tigers. Now in her mid-thirties, she had worked on this operation for three years and enjoyed being outdoors. "Tigers can be mysterious," she said, as the wind blew through her long blond hair. "They can see you, but you can't see them. They like to sneak up on you, too. It is as if they're playing a game with you to see if you can tell where they'll come from."

We motored up to an orange buoy. Raising the aluminum shark cage on the stern with a winch, the crew swung it over the side. The 6-feet-by-12-feet steel cage splashed in the water. As a practice, the crew doesn't drop chum in the water because the sharks have learned to congregate near the sound of the engines. They know that the fishing boats will carve up and throw their bycatch overboard. The engines signal that an easy meal is to be had. For apex predators to survive at the top of the food pyramid, they need such logical reasoning. We just had to wait for the sharks to come to us.

I dropped inside the cage, eager to finally encounter a tiger shark. As I descended underwater, my entire world turned a beautiful monochrome, a beautiful azure vista, dazzling in its simplicity. The islands of Hawaii are a time capsule of the time before overfishing and overdevelopment took their toll on the sea. From inside my cage, a gray mirage appeared. It slinked closer and, as it neared, I could discern its tall dorsal fin. Patrolling the underwater arena with purpose, the shark circled my cage with a graceful push of its tail. Other sharks began to appear and join in: Galápagos sharks. These large sharks are found around the world and are usually 7 to 9 feet long. Dark gray above with a light underbelly, the Galápagos shark has no distinctive markings. I admired the sharks' strength and torpedo shape. They examined me with quick eye movements, curious but without emotion. Not recognizing a source of food, they swam away.

Where was my tiger, though? Ten minutes underwater and nothing. I received the signal to return to the boat. My trip to see the tiger failed, and my quest would have to continue. With patience, I may possibly get to see a tiger eventually. But for now, they are not ready to reveal themselves to me.

The mysterious ambusher, hiding in the blue miasma, probably saw us. It may have wondered what these trespassers were doing but decided we were not worth its time. I can imagine it turning its massive head from us and slowly swimming away to play its role in the ocean. On the boat trip back to shore, I thought about all the ways that tiger sharks make the Hawaiian Islands what they are. Divers that come from all over the world to see the beautiful reefs of the Hawaiian Islands may not realize that the abundant and diverse sea life is the handiwork of the sharks—or that the seagrass is so luxuriant due to the tiger's protection.

Native Hawaiians recognize how the tigers molded

their world, and they came to know them even before the age of telemetry tags. In Hawaiian mythology, the oral myths warned of shark bites in the months of October and November. Meyer was able to show that they were right.[4] Most bites take place in the fall months during the pupping season. Sharks need a tremendous amount of energy to nourish each litter of shark pups during pregnancy. Previous scientific studies have shown that pregnant females of other species of shark are in poor condition in the run-up to giving birth and during the postpartum period. After the females significantly deplete the energy from their livers during pregnancy, they are in greater need for food during these fall months.

Another belief about sharks in Hawaiian mythology is that they are gods. The early Hawaiians understood that tiger sharks are crucial to the health of the planet's oceans, which cover 70 percent of the world's surface. Yet humans have hunted and butchered them and viewed them as creatures only to be feared.

After my dive, I asked Meyer about commercial fishing for sharks beyond the Hawaiian Islands in the Pacific Ocean. He broke down the shark fisheries into two groups. The first group impacting shark populations worldwide consists of the fisheries that are intentionally targeting and capturing sharks. The second group comprises those that are fishing for other fish species, particularly tuna, and catch sharks by accident. "Between these two, undoubtedly, the biggest impact is from the former group, the industrial shark fishing; they are targeting sharks for their fins. And this is a fishery that then takes those fins and sells them to Asian markets, where they're made into shark-fin soup." I asked Meyer how many sharks are being killed for this market. He said, "It's resulting in millions of sharks being removed from the oceans every year, just to supply that demand."

Shark-fin soup is a dish that is associated with affluence in Asia. Historically, the imperial family and court members dined on shark-fin soup during the Song dynasty around 1000 CE. Between the end of the last Qing dynasty and now, shark-fin soup has remained in high demand. Chinese Communist officials have only recently banned shark-fin soup at official functions, more in an effort to end extravagant state spending than as a conservation effort. The long-held and deeply rooted perception that shark-fin soup suggests wealth and status continues. Compounding the issue is the belief in Chinese culture that shark fins contain medicinal properties that can boost sexual potency, clear up bad skin, and prevent heart disease. However, no scientific evidence exists to support this belief. In fact, studies have shown that shark fins contain neurotoxins, and eating shark-fin soup may raise the risk of dementia.[5] Similarly, mercury accumulates in the tissues of top apex predators, and eating shark-fin soup exposes the consumer to mercury poisoning.

The shark fins that end up in soup go through a tortuous path. Countries like Indonesia, Costa Rica, Peru, Ecuador, Spain, and South Africa bring their fins to collection points around the world, and the biggest hub is Hong Kong, which processes 40 percent of all shark fins. The value of each fin depends on three criteria: species, the quality of the fin, and the type of fin. Thick fins—preferably from silky sharks and blue sharks—with discernible veining are considered the highest quality, while the most valuable fin is the lower section of the shark's caudal, or tail, fin, which often goes for between $150 and $300 per pound. Because of a lengthy processing treatment that requires two separate stages of dehydration and skinning, individual fins lose most of their mass. A single kilogram of fin, for instance, yields only 0.1 kilogram of dried shark fin. This is why sharks are finned in such high numbers.

Of the total number of shark fins that enter Hong Kong, 40 percent of the fins processed there are then reexported to other countries, most notably China. The remaining 60 percent of the fins are sold and shipped to dried seafood markets or to tony hotels and restaurants throughout Hong Kong, where the fins are prepared for soup. In addition to silky sharks and blue sharks, vulnerable species like hammerheads and oceanic whitetips routinely end up in this controversial dish.

"Now the problem is that people in Asia as a whole," Meyer explained, "want to serve shark-fin soup, which means the demand for shark-fin soup is continuous." The conservation issue becomes tougher to solve because it's not just about coming up with technical measures that can help reduce the catch of sharks; it's about persuading people not to eat shark-fin soup. Reducing the demand for shark-fin soup is the only way to mitigate the tremendous impact of this industry on shark populations worldwide. Some countries look at the ocean as an inexhaustible supply of whatever fish they want to eat. With millions of sharks dying for soup, it seems that humankind is playing a dangerous game with the planet's top ocean predator.

Many people do not like sharks, and the basis for that view is that sharks are creatures to be feared. But is this fear of being attacked justified? I went in search of the answer to the question: How dangerous are sharks?

Chapter 6
The Shark Attack Files

What are the odds of being attacked and killed by a shark? To answer this question, I visited the keeper of the International Shark Attack File (ISAF), the world's definitive database of shark attacks. For the past twenty years, George Burgess, a professor of ichthyology and marine biology at the University of Florida at Gainesville, has organized the database, a compendium of worldwide shark attack investigations that go back more than four centuries. "We endeavor to investigate every shark attack that has ever occurred worldwide," he told me. "Our earliest attacks go back to the late 1500s." In the first recorded attack, a Portuguese sailor fell off his ship in transit between Portugal and India in 1580. Witnesses on board watched as a shark attacked and killed the sailor.

I interviewed Burgess in his office on campus, where he works out of a brick building. Inside are cavernous rooms lined with rolling storage racks, each one filled with jars stuffed with juvenile nurse sharks, dogfish, and other small sea creatures preserved in formaldehyde. The place is like the Bates Motel for small fish. As I sat with Burgess, his long gray beard and friendly demeanor made me think of George R. R. Martin with a PhD and, based on his prodigious output over his forty-year career, a healthier respect for deadlines. Burgess patiently took me through the file's four-hundred-year history, detailing the specifics of his database with the same seriousness, attention to detail, and scientific rigor he's exhibited since his days as a graduate student at the University of North Carolina.

Under the direction of Burgess, researchers at ISAF record a tremendous amount of information about each attack. They look at the environment, the prevailing conditions, and, of course, the shark. To compile an accurate record, they ask more than two hundred different questions and enter the answers into an electronic database. Like detectives collecting evidence at a crime scene, Burgess and his team analyze data, draw conclusions, and sometimes overturn false accusations. "We have more than six thousand investigations," he explained. "Those investigations look at shark attacks from the perspective of what the humans were doing, what they were wearing, and so forth. Of course, not every investigation turns out to be a shark attack."

For 2018, the worldwide total of unprovoked attacks was sixty-six, below the five-year average of eighty-two incidents between 2013 and 2018. The total number of worldwide fatalities dropped from five in 2017 to four in 2018. Death resulting from a shark attack is highly unusual; a fatality occurs in approximately 5 percent of all attacks, so if a person is bitten, the odds are incredibly high they will survive.

These statistics vary year to year, which can sometimes produce statistical anomalies. In 2015, for instance, the ISAF reported a record-breaking ninety-eight shark attacks around the world, a 20 percent increase over the previous five-year average. Such an increase raised alarms that sharks were becoming a greater threat to humans. In Australia, demands for shark-culling programs increased, and TV newscasters were calling for sharks to be killed wherever surfing contests were to take place. However, 2015 proved to be an outlier. In the three years since, the number of attacks has declined.

I have to admit, I was a little surprised when Burgess told me that the United States is the country with the highest number of attacks. According to him, the

US is ground zero; in any given year, it accounts for about half of the world's shark attacks. The US had thirty-two attacks in 2018, accounting for 50 percent of worldwide attacks.[1]

"The major reason more attacks are in the United States," Burgess explained, "is we've got a very large coastline, two coasts, and we own some islands. And of course lots of people. And we are also a nation of some wealth and means, and so we can afford to spend time at the beach. And of course many areas of the country are heavy tourist areas."

The state of Florida represents approximately 50 percent of the US total, with no other state coming close. After Florida, the other states with the highest number of attacks are South Carolina (three attacks), Hawaii (three), and California (one). Attacks in Florida are high enough to represent approximately 25 percent of worldwide attacks. Florida's long coastline and warm waters, coupled with the state's large population and thriving tourism industry, contribute to its having the most shark attacks of any region in the world. "Everybody in Florida lives only an hour to an hour and a half from the beach, so that means all these Floridians are very heavily oriented towards aquatic recreation," Burgess explained. In addition, the tourist population adds to the number of people visiting the beach. "So, year in and year out, Florida . . . always has a lot of these interactions. And, of course, as Florida goes, so does the United States."

In assessing the grand total of shark-attack-related deaths in the United States, Burgess explained, it's important to look at the averages, because a single year may not be representative. In the first decade of the twenty-first century, for instance, sharks killed an average of 0.4 people per year. "We're also blessed in the United States with very good medical and emergency care and lifesaving abilities on the beaches," he said. "As a result, our death rate is quite low. Of the fifty to fifty-five attacks we have a year, on average, just under one death per year in the entire United States. We do a very good job of making sure that if somebody does get bit, they stay alive." Recent data bear out Burgess's comments; there were no deaths in the United States in 2017, with one death in 2018.

In the United States, a person has a 1 in 265 million chance of being killed by a shark. Compare this fatality rate to the odds of getting killed by a lightning bolt (1 in 218,106) or in an automobile crash (1 in 103) or of dying from an opioid overdose (1 in 96), according to the most recent data.[2] These statistics are based on the assumption that everyone in the United States has an equal chance of being bitten. Technically, this statistic is somewhat skewed, because the true probabilities should be based on the number of people who go into the sea each year, a statistic the ISAF does not yet have data for.

At the same time, other animals kill far more people than sharks do. In fact, sharks are near the bottom of the list of deaths caused by animal attacks. An average of 130 people a year die in the United States following vehicular collisions with deer. Ants kill thirty people per year. Man's so-called best friend, the family dog, is responsible for thirty-four human deaths a year. Even bees kill more people (478) every year than sharks do.[3]

"Most shark attacks are of small scale and, in fact, probably are better referred to as 'bites,'" Burgess noted. "It's not being politically correct; it's just the reality that most of these [shark bites] are the equivalent of a dog bite in severity: piercing wounds or small lacerations that do require some sutures, but not loss of tissue or loss of function. I would say, probably 90 percent or more of all US attacks really are bites. Therefore, anyone bitten has an excellent chance of survival. And these bites are probably cases of mistaken identity in which the shark misinterprets the splashing, particularly of the foot and ankle area in the kicking motion, but sometimes the hands in the swimming motion, as being the movements of normal prey items—fishes."

Australia has the second highest number of shark attacks. In 2018, sharks notched twenty attacks in the waters around Australia, one of which was fatal, a consistent trend from recent years. In spite of the presence of numerous species of sharks including great whites, South Africa has an average of only two attacks per year, and, in 2018, zero fatalities, according to the database. After these countries, attacks are scattered around the globe. Some countries will see a spike in attacks, followed by nothing. In 2016, the French territory of New Caledonia in the South Pacific experienced four attacks, including two fatalities. The territory briefly emerged as "an area of concern," but in 2018, there was only one attack.

The top two species of sharks most dangerous to humans are, not surprisingly, great whites and tigers. Great whites rank first. Close behind is the tiger shark. The ISAF catalogs shark attacks by species. The combined number of attacks and fatalities totals 828 since 1580, and great whites and tigers were responsible for 425, or half, of those attacks, according to the ISAF.[4] Many great white and tiger shark attacks can be attributed to mistaking surfers or swimmers for prey animals like seals and turtles.

The third most dangerous shark species is the bull shark. These sharks are powerfully built and easily recognized by their thick round bodies and tall dorsal fins. Bull sharks, which can be aggressive, are the only shark species that can swim in both salt and fresh water, which increases the likelihood of a human encounter and possible attack. Bull shark attacks have taken place in rivers and inlets around the world, from the Brisbane River in Australia to the Raritan Bay in northern New Jersey. Combined, these three species—great whites, tigers, and bull sharks—are responsible for more bites than any other shark species, primarily because they are easily identifiable. They still only account for less than a third of the total number of shark bites.

Other species, like the blacktip and spinner sharks, are commonly involved in attacks in Florida because the spinner prefers shallow water, where beachgoers are most likely to swim, and the blacktip combs the surf zone for food, often mistaking swimmers and surfers for prey. The attacks by these two species are of the hit-and-run variety, a quick grab and release that can cause minor injuries comparable to a dog bite. The majority of attacks—66 percent of all bites recorded in the ISAF—are from shark species unknown to, or unidentifiable by, victims.

According to Burgess, surfers are the most likely victims of shark attacks: approximately 60 percent of all incidents involve surfers. The United States has 3.3 million surfers, and with only approximately 30 unprovoked attacks on them, the sport is a more than safe pastime. One YouTube video shows a surfer falling off a board, almost landing on a shark, and nothing happens.[5] So why are there any attacks at all? Viewed by a shark from below, a surfboard resembles a seal or turtle or other prey. At the same time, splashing, paddling, and falling off surfboards create disturbances on the water as surfers glide by. "Surfers also make a lot of noise with kicking and falling into the surf," Burgess said, noting that because surfers spend more time in the water than most people, they are more likely to come into contact with sharks.

Less affected recreational user groups include swimmers/waders (30 percent) and snorkelers (6 percent). Scuba divers were the safest, accounting for only 5 percent of the total attacks. One explanation for the low number of attacks might be the low number of scuba divers relative to the larger population. Still, attacks on scuba divers are rare; perhaps being underwater and part of the marine ecosystem is just safer.

While popular opinion holds that sharks are man-eaters hell-bent on devouring unsuspecting swimmers, the data paint a very different story. Provoked shark attacks totaled 34 out of the total 130 worldwide attacks. Surprisingly, people do antagonize sharks by poking them, yanking their tails, and even trying to hitch a ride. It's not surprising, then, that on average, humans cause 25 percent of the total number of attacks. As we dig deeper into these attacks, we see that human behavior can unintentionally increase the odds of a shark bite. One activity that increases the likelihood of an attack is spearfishing, because blood, fish fluids, and electrical disturbances emitted by wounded fish attract sharks. Additionally, spearfishing often puts the fishermen in the water with prey and predator alike. One out of every five victims, according to the database, was attacked with a fish in his or her possession.

It is always difficult to know the shark's motivations behind an attack. Many attacks are likely cases of mistaken identity. While surfers in dark wetsuits on surfboards look like prey items, bathers swimming in turbid waters under cloudy skies also cloak their identity to the sharks. When sharks pursue baitfish toward the shore, they can easily bite a human, thinking they are striking their prey. But shark aggressiveness toward humans is clearly in the minority for explaining attacks.

If sharks were to view humans as prey, we would expect to see certain shark behaviors, and human victims would show certain telltale marks. We would see sharks taking multiple bites, as wolves do. We would also expect to see sharks feed completely on the person who had been attacked. The shark would want to finish off the prey, eating it all, similar to the way a lion eats a wounded water buffalo. Hunger is a powerful driver of behavior, and animals that need food will keep pursuing their prey until they are devoured. According to the ISAF, however, only 5 percent of all attacks examined show multiple strikes (three or more) on the same person. In the vast majority of cases, the shark swam away after only one or two discrete bites. For example, in Perth, Australia, Brian Audas was surfing on his board. A shark popped up and bumped him. He used his arm to push the shark away, but the shark grabbed it and started biting on his arm and moving its head side to side like a dog. Audas said he could feel the flesh tearing away, but as he continued splashing water at the shark, it let go of his arm and swam off. Many cases similar to Audas's are recorded in the ISAF.

If humans were a target for food, we would expect to see substantial evidence of feeding bites on humans. This is not the case. Many of the marks on humans were inflicted by only one set of teeth, mostly the uppers, which would indicate a slashing rather than a biting action. In a typical case from the ISAF, an Australian

Aboriginal diver was bitten on his foot by a shark. He also sustained nasty gashes on the inner side of his right knee. Just as suddenly as it bit the diver, however, the shark swam away. The man's injuries were noteworthy because they were slashes or cuts rather than open wounds. The shark made no attempt to feed on the victim. Examples of this type of attack abound throughout the ISAF. Other sea animals, including dolphins, make similar bite marks, scraping their teeth along the side of their dolphin brethren, either in anger or in play. Like dolphins, sharks often inflict these types of wounds, which suggests that a shark's aggression—against a fellow shark or a human—doesn't always take the form of a bite.

The ISAF database shows that sharks rarely feed on their victims. One man, Frank Logan, was attacked by a white shark in 1968 at Bodega Rock in California's Sonoma County. "I felt something come down on my legs like a giant vise and then a crushing pain in my back and chest," said Logan, who was hunting for abalone when the shark grabbed him by the side, shaking him violently. At that point, Logan went limp and played dead. A few seconds later, he was carried about 10 feet before the shark released him without further aggression. If hunger were the shark's primary motivation for the attack, Logan would have made an easy meal. Even in the highly unusual and tragic case of a lost limb, the shark does not consume the entire victim.

One commonly held view is that a single drop of human blood will precipitate an attack. As we saw in chapter 2, however, modest amounts of blood are not enough to draw the attention of a shark. While it is true that sharks can detect small quantities of substances in the water, a few drops of blood will quickly dissipate in the ocean. In cases where a shark victim was bleeding in the water and more than one shark was in the vicinity, the blood did not draw other sharks. Because sharks are far more likely to home in on low-frequency sounds, such as the thrashing of a wounded fish, a human kicking wildly or paddling vigorously on a surfboard is far more likely to draw the attention of a shark than a few drops of blood are. Even victims who were bleeding profusely, like Frank Logan, were not subsequently attacked after the initial bite: only 4 percent of victims reported being attacked in such a frenzied fashion.

A detailed examination of the ISAF reveals that between 50 and 75 percent of all recorded shark attacks are motivated by something other than hunger or feeding. Sharks are intelligent apex predators, and they learn quickly where to find food sources. Given the millions of swimmers, divers, and beachgoers descending into the oceans worldwide, the sharks would have a field day if they wanted to rely on humans for food. If sharks were truly intent on eating humans, the world's oceans would be nothing short of carnage. The ISAF database goes back to the year 1580, and in the almost half a millennium since its founding, there have been only approximately three thousand unprovoked shark attacks in the world. There is no clearer proof that shark attacks remain a rarity.

Since hunger and feeding are not determining factors in most shark attacks, what then are the sharks' psychological motivations behind them? There are several motivations, and a common one is simply curiosity. Sharks are intelligent animals, and all intelligent animals are naturally curious. When a shark's curiosity is aroused, one way it can gather information is through exploratory bites. Underwater cameras send out an electrical field, which gets the shark's attention. Sharks will sometimes bite a camera to figure out what it is—hardly an act of aggression—as Greg Skomal described happening to his camera in chapter 1. Similarly, many sharks will bump a person to gather more information on the swimmer.

Territorial defense provides another motive. Like any apex predator, a shark will defend its territory, and any violation of that area can elicit a strong response. Rates of attack increased when a person made a sudden entrance into the water. A study of sixty-nine recorded attacks looked at how those people entered the water; 84 percent of those people that were attacked had jumped, dived, or fallen into the water, all of which can be interpreted by a nearby shark as a threat.[6]

During the writing of this book, Burgess retired, leaving the ISAF and the Florida Program for Shark Research in the hands of Gavin Naylor. He classifies some shark bites as a "go away" bite. In one case, John Brothers was scuba diving off Key Biscayne, in Miami, when he came upon the unusual sight of a blacktip shark resting on the sandy bottom near a large stingray. To get a better look, Brothers swam within 20 feet of the pair. His actions seem to have awakened the shark, which swam straight at Brothers with tremendous speed. Brothers braced himself, and the shark swam straight into his midsection and shook his camera. Just as quickly, the shark released the camera and, with the stingray, swam off to deeper water.

Many underwater divers report a change in a shark's behavior or posture before it attacks. Blacktips and gray reef sharks swim erratically before an attack, according to accounts in the ISAF. Some victims noted that the shark's swimming became uneven and sputtering, a drastic change in the shark's usual steady and flowing movement through the water. The shark's tail will quickly lash back and forth, and the shark will seem to be coiled like a spring. James Stewart, a professional underwater photographer, was taking pictures of sharks when his diving buddy noticed a silky shark's head start to swing back and forth. A few seconds later, Stewart was attacked, sustaining a serious bite to his elbow but nothing more. Other observers reported that a shark hunches its back like a wolf, dropping its pectoral fins, before attacking. Two research scientists diving in the Marshall Islands noted this behavior. Rapidly approaching an aggregation of gray sharks, the two divers noticed that a few sharks broke off from the group, dropping their pectoral fins and hunching their backs. Fortunately, the divers were able to leave the area without incident.[7] The shark's behavior in these situations can be systemically interpreted as "threatening pre-attack postures" designed to warn people against further violating the shark's space. These posture changes were reported as being particularly pronounced in a reef environment where the shark's avenues of escape were cut off. Any hiker on a trail knows not to block an animal's path of escape. Swimmers or divers who either didn't see or pick up on these signals from cornered sharks were attacked.

Numerous news articles have reported recently that shark attacks are increasing. I asked Burgess if this were the case. "We're seeing an increase in shark at-tacks," he said, "and have since we started looking at it in 1900. Each decade has had more shark attacks than the previous decade. And the current decade we're working in will have more than the first decade of the twenty-first century." Burgess quickly put to rest, how-ever, the question of whether sharks are getting more aggressive. The increase, he said, "is because we've got more people on the face of the earth every year. As long as the human population continues to rise, and with it comes a concordant increase in aquatic recreation, we can reasonably predict that we will have more shark attacks every year into the future."

More simply stated, more people in the water means more attacks, but the increase in shark-related incidents does not lead to a spike in shark-related fatalities. When media headlines proclaim an increase in the number of shark attacks, as occurred in the summer of 2018 when a twenty-six-year-old man was killed by a great white off Cape Cod, a deeper look at the numbers reveals that shark attacks resulting in fatalities are still rare.

Like the five attacks that captured Americans' atten-tion in the summer of 1916, another famous shark attack occurred in South Australia in August 1963. During a spearfishing competition, Rodney Fox was attacked by a shark. Swimming with his speargun in about 50 feet of water, Fox had fastened a fish float to his diving belt, which allowed him to deposit captured fish there as he swam through the turquoise waters. Suddenly, he sensed a stillness in the water around him. It seemed as if all the sea life had disappeared. Then it happened. A shark hit him on his left side, blasting him with enough force to knock off his face mask and dislodge his speargun from his hand. Fox tried to gouge out the shark's eyes but inadvertently extended his arm into the shark's throat. Realizing his error, he yanked his arm out. The shark's serrated teeth sliced his arm to the bone. When the shark released him, Fox kicked for the surface, feeling the shark body's inches below his flippers. Gulping for breath, Fox grabbed the shark, wrapping his arms and legs around it to prevent it from biting him again. The shark then made a dive, and Fox found himself headed to the bottom of the sea. He released his hold and struggled to the surface, where he was finally able to catch his breath. When he looked below the surface, however, in water red with his own blood, he saw the shark grabbing for his fish float, which was still attached to his belt. The shark dragged Fox and the float back below the surface. As Fox struggled to undo the belt, the shark's teeth snapped the line, which freed Fox to break again for the surface. Fortunately, as he burst through, he saw a boat and immediately started yelling for help.

Fox's injuries were severe. His rib cage, upper stomach, and lungs were exposed to view, as was the white of his radius and ulna. His rib cage was crushed, and one of his lungs was punctured. The boat put to shore, and Fox was placed into a car and whisked to the hospital. Fortuitously, the surgeon on duty had just returned from England, where he had taken a specialized course in chest surgery. The operation lasted four hours. At one point afterward, a priest was summoned to give final rights. Fox was in earshot and shouted, "But I'm a Protestant!" In any event, Fox survived the operation, but he required 462 stitches. In the post-op picture, a circle of black stitches is visible, embedded in his skin from his shoulder down to his hip. Judging by the width of the bite mark, the shark was estimated to be 18 feet long.

It was a miracle that Fox survived. Or as he put it, "I was not ready to go." As severe as the attack was, at no time during the encounter was the shark trying to eat Fox, nor did the amount of blood in the water send the shark into a feeding frenzy. This story is consistent with the data from the ISAF. In a later interview, Fox readily admitted that if the shark had been trying to eat him, the outcome would have been very different.

After his recovery, Fox went on to design and build the first underwater observation cage for diving with the great white shark. For over forty years, he has led major expeditions to film and study his attacker. He arranged and hosted the very first great white shark expedition and, since his attack, has run hundreds of expeditions. Fox currently works to raise public awareness of the plight of all shark species through his dive operation and research foundation, publications, public speaking, and films. In 2010, Fox was nominated for the Indianapolis Prize, the world's largest individual monetary award for animal species conservation.

Another famous attack didn't end as well, unfortunately. This attack took place at Lanikai Beach in Oahu.[8] Just a quarter of a mile off the palm tree–lined beach, reachable with a good swim, are two rock outcroppings called the Mokulua Islands, postcard pictures of the tropical beauty of Hawaii. On December 13, 1958, the sky was clear, although the choppy water was rough. Six friends, ranging in age from nine to fifteen, paddled out to the Mokulua Islands on surfboards and air mattresses. One boy was in a rowboat. After two hours, the six boys began their return trip across the small strait connecting Mokulua with Lanikai Beach. Unbeknownst to them, however, a tiger shark had entered the strait, attracted by the commotion of the boys splashing in the water. Five of the boys moved ahead; the sixth, Billy Weaver, the fifteen-year-old son of a local restaurant owner, struggled on his air mattress. The other five arrived at Lanikai's surf, riding in on a series of waves, which left Billy out in deeper water, alone.

The sound of the ocean breeze was broken by a scream. One of the boys saw a shark at the surface of the water 30 feet away. They glanced at Billy, who was clinging to the air mattress, in great distress. Blood pooled in the water. Billy called out to the boy in the boat for help, but fearing that the boat would be swamped in the surf, the boy turned back to shore. A search for Billy began. Eventually, a helicopter crew from the marine base spotted his body underwater on a reef half a mile south of Lanikai Beach. It was not clear whether Billy had died from loss of blood, drowning, or shock; it was probably a combination of the three. The helicopter crew reported seeing a tiger shark cruising in the area. The shark was estimated to be over 15 feet long.

Billy's situation could have had another outcome. Lifeguards on duty or other adults could have prevented the tragedy. Perhaps if lifeguards were there, they could have gotten Billy out of the water soon after he cried for help. It was an hour and a half before the fire rescue squad arrived on the scene.

This tragedy had a long-lasting effect. In response, the state initiated its Billy Weaver Shark Research and Control Program. Nearly six hundred sharks were caught off Oahu between the months of April and December in 1959; seventy-one were tiger sharks. In an effort to relieve public fears and reduce the risk of shark attacks, the government of Hawaii spent more than $300,000 on shark-control programs. Between 1959 and 1976, six control programs of various size resulted in the killing of 4,668 sharks. Residents also got into the act. Local tourism businesses offered bounties for sharks, and one radio station offered $100 for every shark over 15 feet and $25 for every smaller shark. The argument was simple: killing sharks would result in fewer sharks and thus fewer shark attacks. Subsequent evaluation of the culling programs, however, noted that "shark control programs do not appear to have had measurable effects on the rate of shark attacks in Hawaiian waters. Implementation of large-scale control programs in the future in Hawaii may not be appropriate."[9] The program failed because sharks migrate in and out of particular areas. When one shark is killed, another shark takes its place there.

The deaths of 4,668 tiger sharks did nothing to decrease the number of shark attacks in the state. Hawaii canceled its culling program more than forty years ago. And yet, even today, governments around the world, including the Australian and French governments, regularly call for culling, despite evidence that it does not result in fewer shark attacks. In fact, as Carl Meyer told me, culling "does not do any good and runs the risk of ecosystem-level cascade effects where a general lack of sharks results in boom or bust in populations of species further down the food chain."

The evaluations of the state program from universities and federal government officials were consistent with ISAF's conclusions. There is no link between killing sharks and beach safety. Kim Holland, a shark researcher at the University of Hawaii, told me, "The number of shark attacks has nothing to do with how many sharks are in the water and everything to do with how many people are in the water."

Survivors of shark attacks do not always come away with resentment or anger toward sharks. In fact, the opposite typically occurs. The experience of encountering one of the greatest predators on the planet has touched them. These people now seek the preservation of sharks, not retaliation. Many victims have followed Rodney Fox in seeing sharks as magnificent animals that need protection.

What can you do to lower your odds of being attacked by a shark? Burgess offered a number of suggestions. "There's a good reason why fishes are in schools, birds are in flocks, antelopes are in herds: there's safety in numbers," he explained. And so, when we go into the ocean, we should stay in a group so "we don't become the solitary item that can be picked off by a predator." The ISAF data show that most people who are attacked are swimming alone. Obviously, there are exceptions to the rule, but when a person swims alone, the risk of attack rises. It is best to swim in a group, and the safest position is in the center of the group. In Billy Weaver's case, it was not so much a question of being left behind but rather of being left alone. The behavior of apex predators on land is instructive. Lions and cheetahs select their target after careful study. Once the chase has started, they may run past a slower-moving animal in pursuit of their target. It seems counterintuitive for the predator to bypass a slower animal and continue to chase a faster one, but the explanation is quite simple. Apex predators usually prey on animals that herd in schools. To attack, predators need to single out a specific animal. They cannot be indecisive about which animal they will attack; if they hesitate, they are likely to miss them all. Sharks similarly feed on schooling animals like fish and sea lions, which is why they need to lock on an individual target to capture it. In the unfortunate case of Billy Weaver, the tiger was probably following all six boys as they headed back to shore. As some of the boys entered the surf and Billy fell behind, he became the shark's primary target.

Swimmers should also stay away from fishing piers, and people fishing on the beach. "If you see fish jumping in the water and if seabirds are diving, that usually means there's a school of fishes under them," he said. "If there's fish around, there's almost surely going to be sharks." Splashing creates a lot of noise, which similarly attracts sharks to a specific area. "We're not very graceful swimmers," Burgess said. "Even the best Olympian, Michael Phelps, makes an awful racket when he goes down the pool. He's kicking like mad, he's splashing his arms. That erratic activity is highly attractive to sharks. They're looking for prey items that are in distress. And, of course, in the surf zone where almost all of these attacks occur, the sharks are trying to make a living. They're out there looking for food, and they have to fight the very physical forces that are attractive to us, the breaking surf, the undertow, and all that. Those are things that surfers go out to enjoy; meanwhile, the sharks are trying to make a living. And so, they grab at the first splash they see, and sometimes that first splash is a foot."

Burgess recommends swimming within the break line of the waves. Sharks don't like to swim where the waves are breaking because they might get pushed up on a sandbar or on the beach. Once swimmers go beyond the break line, they are exposed in the open water. Swimmers should be as inconspicuous as possible and should avoid wearing red or yellow bathing suits, because these colors attract attention. And since sharks are nocturnal creatures, swimmers should stay out of the water between dusk and dawn, according to Burgess, because sharks are most active in feeding during these times. In addition, it's harder for sharks to distinguish humans from traditional prey at night.

And if these precautions fail and you are attacked by a shark, what do ISAF researchers recommend? If you are attacked by a shark, they advise a proactive response. Hitting a shark on the nose, ideally with an inanimate object, usually curtails an attack. You should try to get out of the water immediately. If this is not possible, repeated blows to the snout may offer a temporary reprieve, but the result is likely to become increasingly less effective. If a shark does bite, clawing at its eyes and gill openings, two sensitive areas, has proven to be effective, according to the ISAF database. Because sharks respect size and power, one should not act passively if under attack. While the thought of ever having to try to escape a shark attack arouses fear, the odds of being killed by a shark run from quite small to infinitesimal.

Many swimmers forget that humans are an interloper in the oceans. The sea is a wild territory and just as unforgiving as the Serengeti Plain of Africa. Anyone encountering a shark has to be on high alert. While hardly the out-of-control man-eaters they're too often portrayed as, sharks—regardless of species—are not cute and cuddly animals either. They need to be respected.

I think back to my ocean swimming after watching *Jaws*. I should never have let unsubstantiated myths interfere with my enjoyment of the ocean. I wasted a lot of psychic energy worrying about a shark attack. It should no longer be a revolutionary idea that sharks are not out to get us. If sharks were on trial, their defense lawyers would have over four hundred years of evidence supporting their argument that sharks are not a direct danger to humans. Based on the ISAF, it would be hard to find a judge who would convict sharks of harming humans with intent. Yet long ago we placed a guilty verdict on sharks, and so we continue to believe that we have the right to torture and kill them with hooks, guns, and other barbarous means. Old prejudices are hard to eradicate, but we must look at the evidence and make decisions based on rational thinking.

While 99 percent of the time, the ocean is a place of great beauty and carefree entertainment, we must recognize that it is also a place where people can die. When one engages with the ocean, one enters a wilderness, which requires awareness at all times. More than 3,500 people drown every year in the United States, and more than 300 people die in boating accidents every year.[10] These numbers dwarf the number of fatal shark attacks every year. By the same token, anyone who enters the wilderness of the African plains or the Rocky Mountains could encounter apex predators, like lions or bears, which on occasion attack humans. That should not stop people from taking safaris or hiking in the woods. Nor does it mean that we need to kill apex predators. All apex predators, including sharks, need to be respected.

When I was on my flight home to New York after interviewing Burgess, I kept thinking about a comment he made about shark attacks: "Remember that humans are not part of the menu for sharks or any other marine animals. We're not part of that ecosystem. We're a strange object when a shark sees us. And, in fact, most of the time when sharks see humans, they go away."

Maybe at some point in the future, people will examine their fear of sharks and realize it is time to let that emotion go away too.

Chapter 7
The Sex Lives of Sharks

At the southern tip of the United States, more than 100 miles from the continental mainland, lie the Dry Tortugas, a remarkable island group that few people have heard of and even fewer have laid eyes on. After Ponce de León discovered the islands in 1513, he named them in part after the Spanish word for turtles, *tortugas*, in recognition of the many sea turtles populating the surrounding waters. "Dry" was later added to the name to let other mariners know that the islands lacked fresh water. Better known today for its treacherous reefs and its nineteenth-century lighthouse, which was built to guide ships through those reefs, the islands are home to a great number of underwater wrecks— sixteenth-century wooden ships, WWII-era German U-boats—that rival the turtle population, littering the seafloor like discarded scrap in quantities large enough to form one of the richest concentrations of shipwrecks in North America. In the middle of the Garden Key Island, the remains of Fort Jefferson still stand, one of the largest forts ever built by the United States, an expensive and once greatly coveted outpost armed with 15-foot-thick walls and bristling guns that never fired a shot.

No one inhabits the Dry Tortugas; the only human presence comes and goes once a day, courtesy of a passenger ferry that escorts day-trippers from Key West who want to gape at the ruined walls of the fort and snorkel among the island's vibrant coral reefs and kaleidoscopic underwater colors. Beneath the turquoise waters, undulating purple sea fans wave to the passing schools of deep-blue surgeonfish and iridescent parrotfish, while stingrays sweep the sandy bottom for crustaceans. Unique birds, like the rare sooty tern, nest in the quiet of the ocean breezes.

And, of course, there are the sharks.

Because they've been left alone, the shark population around the Dry Tortugas has thrived. Undisturbed, sharks patrol the underwater hills and valleys of the coral reefs and skim over the brain and antler corals. At first blush, it's nothing out of the ordinary. But as I soon discovered, the sharks of the Tortugas were doing something special, something elemental, something the species has been practicing since the beginning of time. The sharks were having sex.

When I began researching shark sex on the internet, two names came up: Jeffrey Carrier, PhD, and Wes Pratt, the world's foremost experts on the sexual behavior of sharks. These two internationally renowned shark sexperts are based in Key West, less than one and a half miles from the Tortugas mating ground.

In the 1970s, Carrier and Pratt, armed with newly minted degrees in marine biology, launched their careers. Sharing a keen interest in sharks, the two men became fast friends and, over the course of their careers, teamed up to publish some of the most remarkable discoveries about sharks. To meet them, I drove four hours west from Miami along Route 1, a scenic stretch of road that runs past translucent green bays and Atlantic shallows. I met Pratt in his office overlooking Summerland Key. There he told me all about his work and everything I ever needed to know about shark sex.

Pratt earned his bachelor's degree in marine fisheries biology and applied for work with the National Marine Fisheries Service (NMFS), the federal agency that is responsible for the stewardship of national marine resources and manages fisheries to promote sustainability. During his job interview, a manager asked Pratt about his area of interest. He responded with one word: "Sharks." Pratt landed the job and, shortly thereafter, shipped off to study large Atlantic sharks. Initially, he focused on the species of sharks with commercial importance: sandbars, makos, and black-tips. He studied their age and growth factors but soon grew fascinated by their reproduction habits. "And who wouldn't be?" he asked. "They're fascinating animals. They reproduce more like terrestrial vertebrates than like fishes. And they've got all kinds of clever organs and systems and strategies."

Pratt showed me photographs of various sharks' anatomies. The male and female sharks are easy to tell apart because the male sharks have two claspers, as Aristotle called them, external organs analogous to a penis. The photo Pratt showed me was of two long, sticklike appendages trailing from the underside of a shark. In the photo of the female, there is no clasper; she has a common vagina, called a "cloaca," suitably designed to receive one of the two claspers. Internally, above the vagina, are paired uteri.

"Sharks reproduce quite differently than your average codfish, your average bony fish, where the female just makes as many as a million eggs per fish and then . . . ejects them at the right time into the environment at a fishing bank or rocky knoll or something. The male just sends sperm into the mass, and that's it. Sperm and eggs—it's all up to them to get together and form these little fragile fish."

Pratt then held up what looked like a stick. "This organ is the male clasper of a 2,400-pound great white shark that was landed in Montauk years ago. And when it's inserted in the female, this whole tip rotates back and splays open like an umbrella to lock it into the common vagina. The tip is very sharp, but she has a thick vaginal pad that receives it."

Sharks, like many animals, are secretive about their mating; they prefer to keep the blinds down, so to speak. During the mating season, sharks stay in schools, but Pratt has shown that, after coital season, male and female sharks separate in something called "sexual segregation." Nurse sharks are unusual because, unlike most large sharks, they mate in shallow water. This allowed Pratt to study them, up close and personal, in the Dry Tortugas.

A top marine scientist at the Mote Marine Laboratory, Pratt has devoted most of his forty-year career to studying shark reproduction and the last twenty-four years trying to puzzle out the nurse shark and its reproductive history because, as he put it, "they will answer my questions and tell me stories better than other sharks." For all of his research, though, Pratt had never visited the Tortugas until he befriended Jeffrey Carrier, nor did he know it was a popular breeding ground for nurse sharks.

After earning his doctoral degree at the University of Miami, Carrier, looking to stay in Florida to research shark populations in the South Atlantic, started an educational program for young students called Sea Camp, where curious children could study marine life in the Keys. Carrier came up with the idea of the program when he noticed a group of small islands 70 miles west of Key West: the Dry Tortugas.

Carrier hopped into his skiff and headed into the Tortugan shallows, where he saw plenty of sharks. They were definitely up to something, Carrier said, but he wasn't quite certain what. Through his binoculars, he saw a bubbling mix of white foam, blue water, and what appeared to be shark tails at the surface. He gunned the engine to get closer, but by the time he arrived, the commotion had come to an end. This pattern occurred, time and time again, and Carrier feared that he was never going to figure out what the sharks were up to.

One time, though, he saw a shark struggling nearby, churning up the water. Carrier assumed the shark had been struck by another boat. He peered down into the water. As he tried to identify the shark, he realized there were actually two sharks, one on top of the other. Anything but injured, the sharks were having sex.

To the best of Carrier's knowledge, humans had never before witnessed such an intimate moment between sharks up close, in the wild. Of course, scientists had witnessed mating in aquariums, which gave them a basic sense of sharks' reproductive behaviors, but some of the most basic questions remained unanswered. "Most of the studies in those days inferred reproductive behavior," Pratt said. "This was back in the late seventies, and we didn't really do anything with it at that point."

Carrier invited Pratt to the mating area around the Dry Tortugas. Their inaugural trip was a bust, but Carrier invited Pratt back the following year. This time their luck turned; they witnessed fifty mating events. Sharks twisting and turning against each other, sending up white water in the light green shallows. "The first time we jumped in the water, we absolutely had no idea what to expect," Carrier said. "We were in about three or four feet deep, and when we got in the water and swam through this cloud of silt that had been stirred up, we found that we weren't face-to-face with two sharks. Rather, we were face-to-face with about eight sharks."

Outnumbered four to one, the odds were not in the

young scientists' favor. But they quickly realized, given the act the sharks were engaged in, that the sharks were a lot more interested in each other than in the two of them.

Studying sharks in the act, Pratt observed that the males have developed cooperative behavior. Together, male sharks work to get a female out into deeper water. To latch onto a female during sex, a male shark uses its teeth for leverage. Pratt showed me a picture of a female blue shark he snapped during a cage dive. The shark is fairly small, probably about 5 or 6 feet long, not the most impressive of species. What is remarkable about the shark, though, are numerous bite marks clearly visible on its dorsal fin. "So this is the tough love of the blue shark," he told me, "but one of my findings also was that the female has skin that's up to three times thicker than the skin of the male to accommodate this behavior that they've evolved over the years."

Pratt then talked me through a video of nurse sharks mating that he took on a recent trip to the Tortugas. "The sperm will go into these claspers and be thrust in by hydraulic action by the siphon sacs of the male, more interesting anatomy for those who like that sort of thing, which I do," he said. "And [the sperm] are propelled up through the uteri, and in some species, stored in special modifications of a shell gland up there for fertilization. Not all sharks do this, but many do. The female can then store the sperm and have young when it suits her. When she's ready, and her liver has been regenerated, like a strong battery, she can self-fertilize."

If only human females were as fortunate.

As a species, sharks bear pups three different ways. Sharks that lay eggs are oviparous, a strategy many fish species follow. Birth is straightforward for oviparous sharks. A black, purse-shaped case that contains the pup embryo comes out of the cloaca. The case has hooks that attach to something in the seabed, like kelp or seaweed. After a few weeks, the shark pup hatches out of the case. Beachgoers find the discarded cases on beaches around the world. Other sharks like tigers and great whites are ovoviviparous, which means the eggs hatch inside the female shark, but the pups are born alive. Because these baby sharks are nourished from the nutrients in the egg, they are born without an umbilical cord. The third way sharks give birth is to bear live pups (viviparity) like humans with placentas and umbilical cords, which lets the mother nourish the young through her bloodstream. Large sharks like hammerhead and bull species give birth this way.

Pratt then held up a picture of a yolk sac and yolk stalk on the ova of a shark. "This is a shark embryo.

It's seven or eight days old." He pointed out the various parts of the embryo, "These are gills. This is probably the beginning of an eye there. And then the tail. It is remarkably similar to a human embryo."

I had to agree. They were remarkably similar. I left Pratt's office with a tremendous amount of new information. But as hours turned into days, I couldn't get that shark embryo out of my mind. One thing Pratt told me continued to resonate: "Ontogeny recapitulates phylogeny." All animals, in utero, go through the same stages of development. If you go back far enough, we all have the same common ancestor as the shark. One could argue that here is the foundation of the Buddhist belief that all animals are interconnected and come from the same universal source. After speaking with Pratt, I thought back to my high school biology classes, when my teacher showed pictures of human embryos in various stages of development. Human embryos have fish tails early in development, an uncanny similarity to sharks at the same stage. Pratt told me that we all come from the same basic plan; we only differentiate as species later.

I wanted to see nurse and other sharks in action for myself, especially since the mating grounds were less than a hundred miles away. I drove to see Pratt's friend and research collaborator, Jeffrey Carrier. Carrier works in Key West, and it was a short car ride over to his office, where he and I talked about his experiences tracking the sex lives of sharks before I was to head to the Tortugas on my own.

I helped Carrier untie his 17-foot skiff, and we headed out into Key West's light-green bay, which was so clear that I felt as if we were skimming over liquid glass. We kept an eye out for shadows moving along the bottom. While at the helm of the skiff, Carrier explained the key relationship between mangroves, one of Florida's true native plants, and marine life. Florida is blessed with an estimated 469,000 acres of mangrove forests, which contribute to the overall economic and ecological health of the state's southern coastal zone. Anchored in mud by multiple thick roots, mangroves are one of the few plants to thrive in salty environments. They obtain fresh water from salt water by secreting excess salt through their leaves; some mangroves block the absorption of salt altogether at their roots. Above the roots and water are thickets of branches that can extend 100 feet above the waterline. They have densely packed small green leaves and form rookeries where coastal birds often nest.

In many countries, the mangroves are being uprooted and cleared to make way for fish ponds. In the Philippines, for example, as much as 50 percent of mangrove areas have been converted, primarily to grow shrimp for the worldwide commercial market. I asked Carrier what would happen if the mangroves here in Florida were cut down for fish ponds. He said, "Without healthy mangrove forests, Florida's recreational and commercial fisheries would drastically decline."

The most immediate benefit of a healthy mangrove ecosystem is the breeding, nursing, and nesting grounds for the local fauna and avian populations, in and out of the water, in either mangrove roots or branches. But the mangroves provide benefits beyond that. Mangroves also help recycle nutrients in coastal waters. As the trees shed their leaves, they drop in the water and are broken down by bacteria and fungi. The detritus of minerals and key elements is then available to the food chain, from plankton to aquatic animals. The mangrove roots not only act as physical traps but also provide attachment surfaces for various marine organisms. Many of these attached organisms filter water through their bodies and, in turn, trap and cycle the nutrients back into the system so the trees and vegetation can grow and start the cycle all over again. Moreover, the nearby ecosystems—terrestrial wetlands, salt marshes, and seagrass beds—also benefit. As the tides come in and out, nutrients from the mangroves are swept into the surrounding areas. This means that there is plenty of food available to a multitude of marine life, including large fish such as tarpon, jack, and red drum.

As Carrier and I were boating, we saw many birds such as brown pelicans and roseate spoonbills, which flapped their wings loudly overhead. Carrier pulled back the throttle, letting our boat drift into a small inlet of white and red mangroves. White herons caused their branches to sag. The water was so clear I could see the mangroves' smooth brown roots buried in the muck. Barnacles clung to the bark. Hidden in the dense, intertwined roots, Carrier told me, were oysters, crabs, sponges, anemones, and many other species.

Because the depth was only 5 or 6 feet, I was able to see the ridges snaking across the water's sandy bottom. As the boat glided to a stop, Carrier dropped a small anchor. Without the motor running, the only sound we heard was the water lapping against the skiff's hull. A gentle breeze softened the intensity of the afternoon sun. In only a couple of minutes, we saw something. A baby scalloped hammerhead slipped by, its head moving side to side, hunting among the mangroves. A perfectly miniaturized version of an adult hammerhead, the shark was, at most, 18 inches long. Its sand-colored skin mirrored the sand below, which made viewing it difficult, but I could still make out the shark's unique head, with scallops carved along the front. A few minutes later, a 2-foot-long baby lemon shark swam into the inlet. Startled to see us, it darted back into the mangroves. We were in the center of a shark nursery. Regardless of how sharks are born, shark pups from the moment of birth must learn hunting skills, or they will perish. In the mangroves, baby sharks are safe during their first years of life before they swim off to deeper, more dangerous waters. Without the mangrove forests, sharks, in particular the nurse shark, would be vulnerable. As nurse sharks grow into teenagers, they stay in the mangrove forest, which acts as their playground. "We don't see the big adults in here, the mature adults, seven and eight footers," Carrier told me. "We'll see nurse sharks only up to about six feet."

Animals often display a marked tendency to return to previously visited locations. Natal philopatry, where animals return to their birthplace to breed, is the most common form of such site fidelity. Salmon are well known for their ability to return to the river or creek where they spawned. To date, Carrier has tagged more than a thousand nurse sharks, mostly in the surrounding mangroves. Through his tagging studies, he demonstrated that nurse sharks return to the same area to mate year after year, in much the same way that salmon return to the same river or creek to spawn.

Carrier and Pratt's research has been corroborated by other studies. Studies of lemon sharks in Bimini, Bahamas, also identified site fidelity among the sharks there. For thirty-five years, scientists witnessed female lemon sharks return to the same spot where they were born to drop their pups.

On the way home, we anchored the boat in a shady spot along a copse of mangroves. Wiping his brow, Carrier summed up what he has learned after conducting shark research for over forty years. He has achieved a remarkable recapture rate of 25 percent of the one thousand sharks he's tagged, which makes him comfortable describing nurse sharks as homebodies. These sharks, Carrier said, don't exhibit any aggressive behavior. In fact, he has never heard of an unprovoked attack by a nurse shark in this area.

After my boat excursion with Carrier in the mangroves of Key West, my journey continued to the Dry Tortugas island by ferry. The turquoise water is remarkably clear and warm, and it's the perfect spot to snorkel to see the biodiversity of the island. I can understand why the nurse sharks flourish here, seventy miles from land and undisturbed by humans. Around these waters, they are born, reproduce, and live out their lives like Adam and Eve in the Garden of Eden, as they have for eons. My visit finally came to an end when the ferry had to go back to Key West.

Because I was still curious about how often sharks mate, I wanted to talk with another shark sexpert. Another researcher with similar interests is Jack Musik, who lives in California. A professor emeritus at the Virginia Institute of Marine Science at the College of William and Mary, Musik has studied sharks for more than fifty years. He began his career in 1961 at Sandy Hook Marine Lab in New Jersey. I was able to talk with him by phone about his research. Curious about how often sharks mated and their gestation period, he maintained meticulous records over the years, which have led to some remarkable discoveries. While humans try—sometimes unsuccessfully—to mate as often as possible, sharks prefer to take their time. "Most males will mate every year, but females less so," Musik explained. "Some females mate every year, but in some species, the females might mate every two or three years."

From conception, a shark's gestation period varies, depending on the biology of individual species and, in general, the size of the shark. The bigger the shark, for instance, the longer the gestation period. Some species like great whites have an eleven-month gestation period. Other species have a relatively short gestation period of four or five months. Despite this difference, the gestation period of sharks is more similar to that of primates than that of fish.

The number of pups also varies by species. Tiger sharks rank near the top of the list and typically produce thirty or forty pups at a time; sometimes they rear as many as eighty pups. Tiger sharks are exceptional, though. On average, most other sharks give birth to about a dozen live young. When it comes to sex and breeding, scientists are still uncovering the reproductive habits of sharks like great whites. They rank at the bottom of the list with only four to eight pups at a time, and no one has ever witnessed the birth of a great white.

Wes Pratt has been working with other species of sharks to find out when they reach sexual maturity. "It takes between five and fifteen years for a shark to mature," Pratt said. They mature like humans, in slow stages. A four- to five-year-old juvenile shark weighs only 13 to 15 pounds, growing at the glacial rate of 4 to 5 inches a year. During their youth, nurse sharks attend school in areas like mangroves, where they learn how to hunt. Pratt added, "In the case of nurse sharks, we don't know exactly when they reach sexual maturity. We suspect that they probably are teenagers at fifteen to eighteen, and maybe start reproducing at eighteen to twenty." Larger sharks like tigers and great whites, however, need twenty to twenty-five years to reach sexual maturity. As far as life expectancy goes, the nurse sharks live about as long as humans do. When the nurse shark dies, its body is recycled back into the marine ecosystem.

Sharks need many elements to reach the pinnacle of their development, but one component is crucial, and that is time: it takes decades for an individual like a great white to be able to reproduce. Fish species have followed two paths when it comes to producing the next generation. Some species at the bottom of the food pyramid produce a prodigious number of young. A female codfish, for example, can lay a million eggs. Sharks, like humans, take another path, which is to produce a limited number of offspring with each one having a long developmental period. Typical shark litters range from as few as five to as many as twenty pups, a far cry from the millions of eggs that other fish produce. "In the case of sharks," Pratt told me, "these animals will replace themselves very slowly."

Considering these factors, it became clear to me that sharks can't survive at the current rate. They literally can't produce fast enough to replace the 100 million sharks that are killed annually. Humankind has taken the ocean's apex predator and upended its position, treating the shark as if it had the fecundity of a codfish.

The history of the cod is a sobering reminder of what can happen when governments ignore the status of fish populations. What happened to codfish is a cautionary tale of what can happen to sharks and other fish species.

In the sixteenth century, Europeans had discovered the Grand Banks, an underwater plateau just to the southwest of Newfoundland, Canada, one of the most productive fishing grounds in the world. The Grand Banks soon became the world's largest source of fish, particularly cod. England, Portugal, Netherlands, France, and Spain (mainly the Basques) sent their fishing fleets to North America to catch them. The English thought of the huge schools of cod as "British gold."

As one of the most sought-after fish in the North Atlantic, cod became the economic linchpin of what was called the "triangle trade." The traders in the colonies dried, salted, and shipped the cod to Europe in exchange for European products to sell. Traders then headed down to the Caribbean, where they sold a lower-quality cod, called West India cure, to slave owners in that market.

Historically one of the most productive fishing grounds in the world, the Grand Banks once yielded what was thought to be an inexhaustible supply of cod. It turns out that this perception was incorrect. The supply of cod began to decline after World War II, when Newfoundland dropped its independence and joined Canada. Jurisdiction over Newfoundland's fishery fell under Canada's Department of Fisheries and Oceans in 1949. To promote jobs in the 1960s, the department allowed an increase in the number of fishing trawlers working the Grand Banks. As the trawlers increased, the cod catch soared. In the 1960s, 2 billion pounds of cod were harvested annually from the Grand Banks.

Over time, local fishermen started to notice dwindling catches. They informed local government officials. In 1986, scientists determined that, in order to conserve cod fishing, the catch rate had to be slashed in half. However, even with these new statistics brought to light, allotments remained unchanged. In fact, the Canadian government continued to predict, without any scientific proof, that the population of the species would rebound from its low point in 1975. Overly optimistic, the Fisheries Department set new quotas in the 1980s based on a drastic overestimation of the total stock. That decision was a disaster. The cod population kept plummeting.

By the time the Canadian government finally acted, it was too late. The number of cod had diminished past the point of recovery. In 1992, the government finally closed down cod fishing. Initially, the 1992 moratorium was scheduled to last two years, during which time it was hoped that the northern cod population would recover, and along with it the fishery. The damage done to Newfoundland's coastal ecosystem proved irreversible, however, even after almost thirty years. To this day, the northern cod population has not rebounded, and the cod fishery remains closed. Even though surviving cod continue to produce millions of eggs, ocean predators pick off so many that only a very few cod make it to maturity.

The closure was devastating economically: 35,000 fishermen and fish plant workers lost their jobs. Today, the once-great schools of cod are only 1 percent of their historic spawning biomass, and the 500-year history of cod fishing has come to a bitter end. Diners may see cod on a menu now and then, but the cod served in restaurants is sometimes pollock or another white fish instead of actual Atlantic cod.

If the cod could be decimated, as massive as their biomass once was, any fish species can be wiped out. When it comes to shark populations, they can be just as vulnerable as codfish. Pratt told me that because most sharks give birth to an average of only twelve to eigh-teen pups per female, they are vulnerable to sustained fishing operations.

Pratt's warning made me worry that the sad plight of codfish could happen to sharks—or any fish spe-cies, for that matter. I wanted to find out just what will happen to the marine ecosystem if sharks go the way of cod, so I headed to Australia, where I soon learned about a world, in the middle of the Pacific Ocean, that I never knew existed.

Chapter 8
Bearing Witness

The Pacific Ocean is the largest and deepest of the world's oceans. It covers a staggering 64 million square miles, approximately one-third of the planet's total surface area, and extends from the Arctic Ocean in the north to the Southern Ocean in the south. The Pacific accounts for 46 percent of the earth's water surface and covers an area larger than all of the earth's land.[1]

Due to the Pacific's sheer size, few island governments have the resources to keep a close watch on illegal fishing and what fishermen are doing with sharks. Without the world's eyes watching, fishing fleets can ravage the oceans and various fish species.

Many nongovernmental organizations (NGOs), such as Oceana and Ocean Conservancy, are addressing ocean conservation, but Greenpeace is one of the few with an active presence on the Pacific Ocean. Their ship, the *Rainbow Warrior*, patrols the Pacific and seeks to protect the ocean from abuses. Today Greenpeace is one of the world's most influential NGOs, with offices in more than forty countries around the world. Even though the organization is not old, Greenpeace, which was founded in 1971, has a storied history. In the late 1960s, a group of activists came up with a radical idea. One of them, Jim Bohlen, a US Navy veteran, learned of a form of passive resistance called "bearing witness." The concept is straightforward: the best way to confront objectionable behavior is to establish a presence where the behavior is taking place. In other words, protesters literally bear witness, a nonviolent and peaceful approach consistent with all great leaders from Martin Luther King Jr. to Mahatma Gandhi to Nelson Mandela. The early founders of Greenpeace knew instinctively that violence is ineffective in the end. In *The Monkey Wrench Gang*, the famous novel by Edward Abbey, the protagonists use sabotage to protest environmentally damaging activities in the southwestern United States. Greenpeace chose to go in another direction, following the peaceful philosophy of men like King, who said, "Nonviolence is a powerful and just weapon, which cuts without wounding and ennobles the man who wields it. It is a sword that heals."

In 1971, the US government planned to detonate an underground nuclear weapon in the tectonically unstable island of Amchitka, Alaska, which lies at the end of the Aleutian island chain that stretches from the Alaskan mainland toward Russia. Activists, who were concerned about the test's likely effect, wanted to stop it.

Jim Bohlen's wife, Marie, came up with the idea to sail to Amchitka, and that autumn, a ship filled with a ragtag group of activists took to the frigid waters and sailed toward the island. The US Coast Guard found out about the plan and sent the USCG *Confidence* to intercept them. Unbeknownst to senior officers, however, the Coast Guard crew sympathized with the demonstrators. The crew members composed a letter expressing their support for the activists' cause and presented it to Jim when they boarded the Greenpeace ship and told the activists to turn back.

There was an internal debate among the activists. Some wanted to keep going, while others were worried about what the Coast Guard would do next. In the end, coupled with inclement weather, the letter convinced the crew to return to Canada. It seemed that the voyage had been a failure. However, once they arrived at port, they realized that their trip was anything but. News about their journey and the reported support from the crew of the *Confidence* had generated enormous sym-

pathy for their protest. Not only did they raise awareness, but world opinion had swung in their favor. It was
a public relations coup. The public outcry persuaded
the US government to discontinue the test planned at
Amchitka. Greenpeace had triumphed in the end.

Since then, Greenpeace's activists have continued
to travel the world on their three ships to protect the
oceans. One of those ships, the *Esperanza* documented
unsustainable and illegal tuna fishing in the Pacific
Ocean in 2011. During the campaign, which covered
more than 14,000 miles, the *Esperanza* crew encountered sixty-three fishing vessels and took appropriate
actions, including boarding and inspecting the vessels.
Greenpeace's two other ships—*Arctic Sunrise* and the
Rainbow Warrior—joined the *Esperanza* in the organization's campaign to save the seas, the latter of the
two focusing on tuna fishing and shark poaching in
the Pacific.

I asked Greenpeace officials if I could interview the
crew of the *Rainbow Warrior* to learn what is happening to sharks on the high seas. I was told that there
might be a window to get some time with the officers
when the boat was docked in Busan, South Korea. I
leaped at the chance. Before Greenpeace officials could
change their minds, I was on a flight westbound. A

A great white shark smashes through the waters of False Bay in South Africa. Some great whites can jump 10 feet above the surface. *Chris Fallows*

A baby seal, whose life literally hangs in the balance, scrambles to escape the teeth of a great white in False Bay. *Chris Fallows*

After Steven Spielberg's *Jaws*, local shark tournaments rose in popularity, which resulted in the widespread massacre of sharks. Makos, threshers, and other species were killed—all in the name of sport. Pictured here, circa 1975, is a great white shark. *NOAA*

After the tournament, the sharks are cut up and their individual body parts are often sent to landfills. Here, a shark's still-beating heart sits on the front loader of a tractor in Montauk, New York. *William McKeever*

A mako killed in a recent shark tournament in Montauk. Desperate to spit out the fisherman's hook, the shark vomited out its stomach. *William McKeever*

Combined with its intelligence and toughness, a mako's speed makes it a formidable underwater hunter—though no human has ever been killed by a mako. *Howard Chen/iStock*

Perched at the extreme ends of its head, a hammerhead's eyes afford it a 360-degree panoramic view of its environs. In addition to its trademark eyes, the hammerhead's sensory abilities allow it to detect prey buried in the sand. *Duncan Brake*

Schooling hammerheads in Costa Rica. Dominant females work their way into the center of the school by pushing away females lower in the hierarchy. Males then fight their way into the center to mate with the top females. Over the past forty years, the hammerhead population has declined by 90 percent in the North Atlantic and Gulf of Mexico. *Janos Rautonen/Shutterstock*

The hammerhead's tall dorsal fin creates tremendous lift, which allows the shark to swim on its side, conserving energy. *Duncan Brake*

Two tiger sharks on patrol in the Bahamas. Although the tiger's stripes fade as the shark ages, the stripes are still visible on both tigers here. Tigers protect areas of seagrass by keeping away turtles and dugongs, which feast on the flowering plants. Tigers also keep the ocean clean by eating man's discarded junk, including tires, license plates, and other trash. *Duncan Brake*

Lemon sharks swimming in the mangroves of Key West, Florida, which serve as nurseries for these sharks before they head out into deep water as adults. Despite the popular misconception of being solitary, lemon sharks are social animals, regularly teaming up with partners for protection, playtime, and lessons on how to hunt. *Duncan Brake*

The author speaking with Joe Krone (*left*), a surfer attacked by a great white, in Mossel Bay, South Africa. The great white took a bite out of his board. Like many people who are attacked by a shark, Krone said he has more respect for sharks as a result of the encounter. *William McKeever*

The author diving with Caribbean reef sharks off Grand Bahama Island. These sharks help maintain the vital balance of the ecosystem of the reef. Where sharks are overfished, as they are in parts of the Caribbean, reef habitats suffer. *William McKeever*

Inside the office of Mark the Shark, a legendary Miami-based charter boat captain—and self-proclaimed "last great shark hunter"—where shark jaws dangle from the ceiling. Customers hire charter boats to live the *Jaws*-like myth of battling man-eating sharks. In reality, most attacks result in injuries about as serious as a dog bite. *William McKeever*

Not all sharks are big and scary. A leopard catshark, named after its spots, peers out from his tank. *South Africa Shark Conservancy*

A scientist at the South Africa Shark Conservancy in Hermanus, South Africa, handles a pajama shark, a species endemic only to that country. Notice the long horizontal stripes along its body. *South Africa Shark Conservancy*

Some sharks fit easily into a human hand. This baby catshark is just beginning its life. *South Africa Shark Conservancy*

Longline fishermen in the South Pacific hauling in a blue shark, the species most commonly caught by longline vessels. Because the demand for shark fins remains so high throughout Asia—China, in particular—sharks are caught and finned, then thrown back into the ocean to suffocate to death. *Greenpeace*

Shark fins stored in a freezer compartment. By keeping only the fins of a shark, fishermen save room on board for other fish, such as tuna. A fishing vessel's catch is often transshipped to reefer vessels, which allows the ship, and its crew, to remain at sea for extended periods of time. *Greenpeace*

Fishermen, who rely on the sale of shark fins in lieu of a regular paycheck, often work 18 to 20 hours a day. Finally allowed to sleep, they fall, exhausted, into their crowded bunks. *Greenpeace*

Shark fins deposited at a port facility in Hong Kong, which serves as the main depot for more than 40 percent of the fins that end up in shark-fin soup. While consumption of shark-fin soup is falling somewhat in China, the dish is increasingly being served in other Asian countries like Vietnam, Thailand, and Indonesia. *Alex Hofford/Greenpeace*

Bearing witness, Greenpeace's iconic *Rainbow Warrior* inspects fishing vessels in the Pacific Ocean to make sure they are fishing legally. Illegal, unregulated, and unreported (IUU) fishing is a major challenge to sustainable fishing around the world. *Greenpeace*

brief stopover in Taiwan gave me enough time to visit the country's busiest seaport in the north.

Taiwan's fishing fleet, one of the largest in the world, ranks third internationally as catchers of tuna, hauling in 325,000 metric tons of the fish per year. At the port city in Donggang in the west central section of the country, I saw a huge number of fishing boats at anchor. These vessels were a massive 200 feet in length with black-painted hulls. Each vessel had cranes on deck that could be mistaken for gun turrets. I counted twenty of these fishing battleships in a single row. The port reminded me of the massive Port Newark–Elizabeth Marine Terminal, except instead of container ships, the port was filled with fishing vessels.

In another area of the port, I saw a totally different kind of ship, not the huge iron-plated battleship types. These ships were smaller, anywhere from 70 to 80 feet long, with only a long wheelhouse at the top where the captain pilots the boat. The wooden decks were old and seemed dingy. Some boats offset the dull planks with painted stripes along the wheelhouse. All the railings were wood. The vessels had a saddleback appearance like an old horse, with a sagging middle and upturned ends.

Taiwanese fishing vessels have been accused of illegal and ecologically harmful fishing practices.[2] Allegations include fishing in the territorial waters of other countries without authorization, overfishing, and shark-finning—charges that cast a negative light on the country, which prides itself as being a responsible global citizen. Leery of drawing attention to myself, I tried to walk discreetly among the vessels as I peered into the ships. Could these small boats hold enough sharks and tuna to make a profitable trip, or was I missing important pieces of information?

And another question: Where was the crew? The Taiwanese fishing industry employs 317,000, a number that doesn't account for the tens of thousands of migrants working in the industry.[3] Some researchers estimate that as many as 160,000 migrants currently man Taiwanese fishing vessels.[4] What sorts of conditions exist on these boats? I noted that there should be more men working around the docks, but the boats were mostly deserted. Why? Unfortunately, I would have to wait to answer these questions.

My flight touched down in Seoul, a modern city that looks as if it would fit in just as comfortably along the East Coast of the United States. Giant skyscrapers pierce the skyline, and the wide boulevards are jammed with traffic.

The *Rainbow Warrior* was docked in Busan, in the

southeastern area of Korea. After what seemed like a never-ending train trip, I stood on a wharf and stared at the *Rainbow Warrior*. Greenpeace named the ship after a North American Cree Indian prophecy: "When the world is sick and dying, the people will rise up like Warriors of the Rainbow." Three ships christened the *Rainbow Warrior* have borne witness around the world since the late 1970s. The ship I was looking at was purpose-built and completed in 2011. I was surprised at the size of the current *Rainbow Warrior*. It is 200 feet long and has two masts, both of which extend a remarkable 180 feet above the deck. In addition, the ship has thirty berths, more than enough to accommodate the crew needed to wage effective campaigns. The hull is painted green and has a rainbow on the bow, the ship's world-famous insignia. Even though the ship sails primarily under wind power, to reduce its reliance on fuel, the new *Rainbow Warrior* is as fast as many industrial vessels; its action boats can be deployed in minutes. A helicopter landing pad similarly allows for immediate engagement, affording the captain and crew a vital eye-in-the-sky advantage when tracking illegal fishing operations. A Zodiac boat rests on a portside crane, ready to launch.

As I boarded the ship, I was inspired by all the great work it has done and the dangers its crew has stared down. It felt like an honor to stand on the vessel and consider the brave crew and their campaigns. The first *Rainbow Warrior* withstood Russian grenade launchers when Greenpeace's Zodiacs were protecting whales. Then the vessel was attacked on July 10, 1985, in Auckland, New Zealand, when bearing witness against and protesting French nuclear testing in the Moruroa atoll. At close to midnight on that night, Captain Pete Willcox and most of his crew were sleeping, while a few others, including the photographer Fernando Pereira, chatted in the mess, drinking the ship's last two bottles of beer. Suddenly, the lights went out, followed by a sharp crack of breaking glass and the sudden roar of water.

Those already on deck scrambled up the ladder or leaped to safety on the wharf. Within minutes, nearly everyone was off the ship and watching the steel masts tilting toward them. "I stood there looking at the boat with all of these bubbles coming out of it," Captain Willcox later recalled. That's when a crew member told him that Pereira was still on the boat. The Portuguese-born photographer had joined the crew of the *Rainbow Warrior* to document the French nuclear testing and share his photographs with the world. He was caught in a rush of water that night and drowned. He had just celebrated his thirty-fifth birthday.

The *Rainbow Warrior* wasn't hit by a boat, of course. In an attempt to "neutralize" the ship ahead of its planned protest, French secret service agents in diving gear had attached two packets of plastic-wrapped explosives to the ship, one by the propeller and one to the outer wall of the engine room. Initially, the French government denied all knowledge of the operation, but the evidence of its involvement was overwhelming. Eventually, the prime minister appeared on television, chastened, and told a shocked public: "Agents of the DGSE [French secret service] sank this boat. They acted on orders."

Only two agents ever stood trial. Dominique Prieur and Alain Mafart, who had posed as Swiss tourists, pleaded guilty to charges of manslaughter, earning sentences of ten and seven years. A settlement negotiated by the United Nations allowed their transfer to the Hao atoll, a French military base in French Polynesia, where they served less than two years.

Thirty years later, a French secret service diver named Jean-Luc Kister apologized for attaching a mine to the ship's hull. "I have the weight of an innocent man's death on my conscience," he said. "It's time, I believe, for me to express my profound regret and my apologies."

While it is unusual for crew members to be killed,

Greenpeace work can be dangerous. Many fishing boats routinely ignore international law, fish without a license, and don't want to be caught. In many incidents, the fishing vessels hurl epithets and, more menacingly, objects at the boat and crew. Warning gunshots are often fired. Once, an individual from a fishing vessel's crew hurled objects at the Greenpeace helicopter, smashing its windows. In 2016, Greenpeace painted STOP ILLEGAL FISHING across the hull of a Taiwanese fishing vessel. Furious, the captain sent his men out in small crafts to attack. They rammed the Greenpeace Zodiacs, and crew members attacked the activists with metal rods. One activist was gouged with a rusty drum hook.

When a ship is in national waters, local authorities can protect their jurisdictions by boarding the ship to make sure the vessel has the appropriate authorizations and to check its logbooks. Unfortunately, many coastal states are too poor to provide the necessary ships to patrol and inspect every fishing vessel, but inspectors can take advantage of the *Rainbow Warrior* to monitor vessels in their waters.

I was given a grand tour of the *Rainbow Warrior*. Below deck, I met the cook in the ship's galley, where plenty of refrigerators held enough food to maintain the crew over long voyages around the world. The cook was busy preparing a lunch of tossed salad for the crew, many of whom were chatting amicably in the mess hall. Behind the mess hall was a library with a wide selection of books and periodicals. The cabins looked comfortable. Each one was furnished with a small bedside reading desk, which made the 12-by-18-foot room feel like a college dormitory.

Finally, I met Hattie, one of the first female captains of the *Rainbow Warrior*. Like Madonna and Beyoncé, Hattie goes by a single name. Sailing is in her veins; she started in the Netherlands when she was fourteen years old. She joined Greenpeace in 1992, after sailing charters for tourists. "I see Greenpeace as the voice of people, the voice of the environment," she told me in a heavy Dutch accent. "Bearing witness is one of the most important things we do. We need to show the world footage, pictures, of illegal fishing because we need to do something about it." To illustrate her point, she took out her computer and pulled up photographs of mutilated sharks and other apex predators, like swordfish and tuna.

Tuna migrate hundreds of thousands of miles over their lifetimes. More often than not, tuna vessels fish in international waters, which makes regulation difficult.

The current mission of the *Rainbow Warrior* is to raise awareness about the tuna-fishing industry and stop illegal and unreported fishing. Tuna, like sharks, have a large number of subspecies and can range in size from just a few pounds up to more than 1,500 pounds. An adult bluefin can weigh more than a polar bear. Tuna have sleek, torpedo-shaped bodies and, like sharks, are apex predators. The ancient Greeks admired the fish, which were a visible part of the Mediterranean culture for much of recorded history. Because sharks hunt tuna, wherever tuna congregate sharks are sure to be nearby.

Today, coastal countries around the world rely on tuna for food and jobs. The tuna fishery is the largest fishery in the world and is worth $40 billion.[5] The canned tuna market accounts for $30 billion. While the price per metric ton is modest, the total tonnage is enormous. Currently, almost 80 percent of all landed tuna goes to canneries for processing. The other $10 billion is generated from various segments. Tuna-based cat and fish meal, for instance, is worth approximately $1 billion. The Pacific Ocean is home to nearly three-quarters of all tuna landings.

Karli Thomas works on the *Rainbow Warrior* as the tuna campaign manager. A native of New Zealand, she has worked for Greenpeace for ten years. She has long, wavy brown hair, and her blue eyes show a keen intelligence. "My goal is to expose what's going on in the world's largest tuna fishery," she told me. "The central and western Pacific Ocean is where 70 percent of the world's tuna comes from." Americans eat on average 2.1 pounds of tuna per person per year,[6] enough to make tuna the country's second most popular seafood of choice, behind shrimp. "Most people have no idea about how [tuna is] caught, what's the status of the tuna stocks, and the plights of the people that are catching their tuna. And that's what we're trying to expose with our ship."

China, Indonesia, Thailand, Korea, and other Asian countries have huge fleets, ranging from small artisanal-scale vessels operating in coastal waters to medium- or large-scale domestic vessels operating within national waters and on the high seas. In addition, many countries possess large-scale distant-water foreign vessels capable of operating far from their home base in any ocean. More than forty countries currently host tuna-processing industries.[7]

Several fishing methods are used to catch tuna: pole and line, purse seining, and long-lining. Just as its name suggests, pole and line involves one fisherman catching one fish on one pole at a time. While environmentally friendly, companies seek out more efficient methods to catch tuna. Industrial purse seine gear, which captures the majority of the world's tuna stock, encircles schools of tuna with large nets. The top of the net is mounted on a float line, while the bottom of the net is fashioned to a lead line, which usually consists of steel chains and rings known as "purse rings." Purse seine nets can measure a mile long and run 200 yards deep, the length of two football fields.

Longline fishing accounts for the next largest catch of tuna and is a remarkably simple method of capture. The crew sets a single line from the stern of the boat. Attached to the line, at regular intervals of 10 feet, are shorter secondary lines. To this one line, fishermen attach thousands of baited hooks. "So, some of these vessels, the ship itself, can be relatively small but the lines that they set out are huge," Thomas said. "And we've been onboard a vessel that had already set more than 100 miles of lines and was still setting." She noted that the tuna longline industry is "an industry out of control. More than 3,500 vessels are authorized for tuna fishing in the Pacific, and yet it is well known there are many more that aren't authorized."

When one considers unauthorized vessels, there are probably more than 5,000 vessels hunting tuna this way. That's up to half a million miles of longline stretching across the Pacific at any given time. "A big part of the issue of long-lining," Thomas pointed out, "is the sheer number of vessels out there, the size of the lines that they're setting, the number of hooks that are on there."

Putting a baited hook in the water is an indiscriminate method of catching a target species. Anything can get caught on that line. While longline fishing allows boats to catch tuna as cheaply as possible, the unintended consequences are that the method sucks the life out of the ocean, like a vacuum cleaner. The method has global impacts on fish, seabirds, and marine turtles. Greenpeace estimates that 300,000 turtles and 160,000 seabirds are killed each year along with other discarded fish species too numerous to mention. Their remains are dumped overboard to sink to the depths. The impact on various endangered species, such as the leatherback turtle and the albatross, is inestimable.

Because the sharks follow the tuna, it is inevitable that sharks also fall prey to longline fishing catches. "The actual catch rate of sharks with longline fishing can be phenomenally high; more than 50 percent of the catch is actually sharks," Thomas said. "Millions of sharks get caught every year in longline fisheries." Greenpeace estimates that, for every ten tuna caught, five sharks are killed.

Thomas and I continued our conversation outside, standing along the gunwale of the *Rainbow Warrior*.

A stiff breeze—a welcome respite from the powerful sun—carried the scent of the sea air. Thomas's ponytail swayed gently. Her blue eyes, though, continued to blaze intensely as she related her firsthand experiences. With the blue sea as her backdrop, she described with fulsome details the savagery that befalls sharks on the high seas.

Blue sharks are the most common casualty since, as scavengers, they regularly go after the baited hooks. Hauled on board, a shark can inflict damage, so the fishermen immobilize it by stabbing it repeatedly in the head or gills. There is no mercy now. When the shark is no longer a threat, the fishermen start hacking it apart. Most shark species have as many as seven fins—two pectoral and dorsal fins plus an anal and pelvic fin on the underbelly, and of course the tail. None of the fins is allowed to go to waste. The easiest to remove are the pectoral, or side, fins, followed by the shark's trademark dorsal fin. The fishermen saw back and forth, exposing the white connective tissue and raw red muscle beneath the shark's skin. Once a fin comes off, a fisherman tosses it into the fin pile. Some fins are small, like the pelvic and anal fins, which are located toward the tail, but, regardless of size, they are all useful for making Chinese shark-fin soup. The last fin to come off is usually the caudal, or tail, fin. Because the caudal fin is the largest, severing it requires more time and effort. Inside the caudal fin, which powers the shark, is a rich supply of blood, which gushes out over the deck.

In many cases, even after this brutal removal of all its fins, the shark is still alive, barbarically reduced to a cylindrical stump.

When the fishermen are done chopping up the sharks, they stuff the fins into bags, which are then stored in freezers located deep in the ship's hulls. The loot will bring in over $300 a pound. To dispose of the shark, the fishermen unceremoniously dump it overboard. Unable to swim, the shark sinks to the bottom of the ocean. "The amazing cruelty is that without its fins, it's simply going to die a painful and slow death in the ocean," Thomas said. "I am angry at how the sharks are made to suffer and face the pain and agony of such an end. Some of the shark species have had huge declines; the oceanic whitetip in the Pacific region has declined by more than 90 percent. So some of these species are in huge trouble."

A first mate named Fernando Martin told me about a recent experience aboard a fishing vessel. He and three other members of Greenpeace, including a regional inspector, confronted the ship's captain about his catch. "He kept two notebooks—one to show to authorities the right amount of catch he was allowed, and the other notebook was for the company and for himself," said Martin, a handsome Spaniard who's been with Greenpeace since 1995. The catch didn't match the records. Fernando and the others decided to search the boat. "We were inspecting the holds. In the beginning, [the captain] just led us to inspect one of the holds. But then we found a new door. He didn't want us to go in that door."

Like in an Alfred Hitchcock movie in which the protagonist is about to do something the audience knows he shouldn't, Fernando opened the door. "What we found inside were several bags full of shark fins." Technically, fishermen are allowed to catch specific shark species. However, some countries require that entire shark bodies with fins attached be brought back to shore. For example, according to the 2010 Shark Conservation Act, all sharks caught in US waters must be brought to shore with their fins untouched. But keeping the bodies as well as the fins takes up valuable space. As I noted when I was in Taiwan, these ships have limited cargo space, so they cannot bring back the shark's body. Fernando estimated that they discovered between 60 and 70 kilos (132 to 154 pounds) of shark fins. "We imagined that they just cut the fin and throw the body to the water while the shark is still alive."

Thomas's experiences were even more shocking. During one inspection, she checked all the freezer holds. In the last freezer hold she found three sacks with more than six hundred fins from various shark species. The inspectors spread the fins out on deck and took photos of them, but later they still couldn't identify all of the specific species by the detached fins alone. Of the six hundred fins they photographed, they were able to identify only four species: blue sharks, makos, scalloped hammerheads, and silky sharks. Only blue sharks appeared in the captain's logbook. Scalloped hammerheads are an endangered species, and because of their declining stock, they are protected against fishing of any kind around the world. To me, silky sharks are one of the most beautiful shark species. They look like bronzed torpedoes with their distinctive small dorsal fin and long pectoral fins, but they are rarely seen by the public because, as pelagic creatures, they live in the high seas. However, those long pectoral fins and their presence in the middle of the Pacific Ocean means that they are particularly vulnerable, so it is not surprising that their populations are declining. As a result, at the seventeenth Conference of the Parties to the Convention on International Trade in Endangered Species (CITES), silky sharks were added to a list that allows the parts to be traded worldwide only if they are documented to be legally and sustainably sourced. While not a ban, the measure does make it more difficult to exploit the species.

When Thomas asked the captain about the missing shark carcasses, he admitted that he had transferred part of his catch. "That transfer was illegal," she told me, "because he didn't have the requisite paperwork to do that." Thomas and the other inspectors reported the captain to regional authorities and to Taiwan's Fisheries Agency, which sent a patrol boat to check the vessel. On board, they found only one hundred shark fins, far fewer than the six hundred fins Thomas had found.

On another search in international waters, Thomas came across a mixed fleet from Japan, Korea, and Taiwan. After examining a dozen or so vessels, she discovered that one wasn't in possession of a fishing license for that region. She and the team boarded the ship to investigate. "The first thing that tipped us off that something was dodgy about this vessel was the fact that its logbook really didn't line up for the amount of time that it had been at sea already, which was more than two months."

How can a ship stay at sea for well over two months? The answer is transshipment, a devilishly clever way to stay at sea indefinitely. On the open sea, ships transfer their catch to other vessels, which then take the catch back to port and to market. Huge vessels called "reefers" are designed to take the catch of various fishing vessels into their gargantuan freezer holds. After unloading their catch and getting fuel and food, the fishing vessels can continue fishing and stay out at sea. This method sounds like an efficient way of doing business, but it also wraps a black cloak around fishing atrocities.

"It means that the fishing vessel stays out of reach of inspection," Thomas said. "It means that the fishing crew don't even come into port—they have no chance to get off the ship if they're being exploited. And it means that legal catch can be mixed with illegal catch. And when [the catch] comes to shore, there's no way of knowing which vessel it came from."

One Greenpeace video taken of a longline vessel shows the horror these boats unleash on marine life. In the video, the vessel dropped a 150-mile longline from its rusted stern into the deep blue Pacific Ocean. The line disappeared into the whitewashed wake of the ship.

The next day, the vessel's winches hauled in the line with a cornucopia of sea life attached. A huge leatherback turtle, a hook sticking out of its beak, was hauled in, along with dead gulls, albatrosses, and other seabirds that drowned after getting ensnared in the line. A striped marlin was the next fish to be hauled in. Its huge rapier bill was 5 feet long. Gasping for oxygen, the marlin slid into the discard pile like an old shoe. At last, the targeted prey, a yellowfin tuna, appeared on the line. Crew members gathered the tuna to throw it in the hold. A slender blue shark was next in line, still alive. The line could have been cut to let the shark go, but the men knew to gather it in for its valuable fins.

With a 3-foot-long blade, a crew member started stabbing the shark. Its tail flailing, the shark futilely tried to escape while its blood spattered over an already slippery deck. The fisherman sliced off the shark's dorsal fin and then methodically chopped off its pectoral and tail fins, reducing the shark to a bloodied stub of its former self. The fisherman threw the mutilated shark back in the ocean, where it twitched from side to side as it sank to the seafloor.

Everywhere these longline vessels travel, they bring death. The damage to the high seas is devastating. As many as 100 million sharks a year are killed, along with tens of thousands of other animals. Every year, the total shark deaths are equal to one thousand times the American casualties in World War I. When Americans eat canned tuna, they do not realize the destruction of the ocean that their meal represents. Imagine if producing a single hamburger required butchers to kill not only the cow but all the other barnyard animals, too.

When I first saw the ships in Taiwan, my picture of the tuna-fishing industry was incomplete, but now I can piece together the giant jigsaw puzzle. The process is simple: The smaller fishing vessels go out and catch the tuna and the sharks. They fin the sharks, stash the fins in bags, and dump the bodies overboard to save room on the boat. Next, they cram the tuna into the ship's hold. When the seams are bursting, they run to the reefer ship for a transshipment of the tuna and the shark fins. Now the vessel is ready to repeat the operation—again and again. For these boats and their crew, time stands still at sea.

When my experience on the *Rainbow Warrior* was over, I returned home to New York, where I took the time to contemplate everything that I learned. I couldn't help but think about what my friend and fellow shark enthusiast Duncan Brake said at Shark-Con earlier in the year. Filming a recent excursion between the Falkland Islands and Antarctica, he experienced an odd occurrence. Five miles offshore, he and his shipmates spotted something bobbing in the water. At first, they thought it was a buoy. As they got closer, though, they realized it was a body, a human body. Brake and his shipmates brought the body onboard and transported it back to port in the Falklands. While they couldn't identify the deceased, they knew he was Asian and had died of hypothermia. I wondered how an Asian man had come to drown in the South Atlantic, so far from home. The answer was perhaps related to the fishing industry. I know Asian fleets venture around the world for fish and sometimes enter the South Atlantic. Why are these fishermen so desperate to escape? When they see land, they risk swimming for it. They don't worry about the distance or the water temperature. Unfortunately, they don't realize that when the waters are frigid, it will take only twenty to thirty minutes to get hypothermia.

What is happening on these boats?

Chapter 9
Human Trafficking at Sea

Learning that some fishermen will risk death to escape from fishing boats made me realize something is out of the ordinary. What could be forcing the men to stay on the boats? Could human trafficking be taking place in the fishing industry? Since fishing is a worldwide industry, a number of nongovernmental organizations are advocating for workers' rights on the high seas. While researching this book, I traveled to London to visit with David Hammond, who at the time was the first chief executive officer of Human Rights at Sea (HRAS), an independent UK-based maritime human rights charity, to learn more about working conditions on commercial fishing boats.

A former officer and lawyer in the British Royal Marine Commandos, Hammond has a strong no-nonsense presence. His status as an international human rights barrister and seafarer early in his career also helps with his cause. One has the expectation that he could physically stand up against anyone, yet underneath the tough exterior is a man with a deep concern for the weak and marginalized.

The founding principle of HRAS is that human rights apply equally at sea as they do on land. As the leading independent maritime human rights platform, the organization has published a number of high-profile papers documenting abuses at sea and has launched a number of investigations and provided support for fishermen formerly held as slaves. Today, they have ongoing case studies of human rights abuse from Fiji to Iran, United Arab Emirates, and the central Mediterranean area. The pattern of such abuse in the maritime environment is emerging as being systemic, but yet significantly underreported on a global scale. No one knows the true scale of the issue.

Hammond shared with me the chilling stories of several young men who were held captive at sea. For their safety, their names have been withheld, but all of the men discussed here are now trying to get back to their home countries, which is a real challenge without money or a passport.

One man was kidnapped at age seventeen while trying to top off his telephone credit card in a local shop. He said, "A man grabbed me and put his hand over my mouth. I ended up in the boat. My parents are still wondering where I am."

An eighteen-year-old from northern Bangladesh was seeking work in Dhaka, the capital of the country. An elderly man offered him a job that paid $6 a day. The two traveled to a small remote house, where the young man was bound and drugged. When he woke up, he found himself on a boat. Once at sea, he was repeatedly beaten.

A third man, an electrician, was kidnapped by a gang of men at a house where he was scheduled to do repair work. As soon as he entered the house, the men grabbed him and knocked him out with chloroform. Later, the electrician woke up on a fishing boat, miles out at sea. To this day, he is still trying to return home to his mother. "Since I was kidnapped, she has no one to give her medicine," he said.

Hammond, who retired as CEO of HRAS in October 2017, though he still works with the charity's board of trustees, described a typical fishing boat as a floating factory floor filled with industrial-size spinning winches, swinging fishing hooks, and legions of roaches and rats. "If you don't know what you're doing and you put your hand in the wrong place," he told me, "it will get lopped off. Now imagine all those machines running all the time, in the middle of the night. Now put that factory on a floor that goes up and down 15 feet, continuously, all the time, and sways side to side. Now make the floor an ice skating rink from fish parts and blood, and now put forty people that are barefoot on there and have water splashing over the side every two minutes. That's the boat deck."

In one case, a man lost four fingers in an accident. He was told to swim to another boat 50 yards away, according to Hammond, which he did, holding his maimed hand above the water. One trafficked fisherman told Hammond that he endured regular beatings by the captain, who used to whip him with the tail of a stingray. Another fisherman recalled a different captain telling him, "I killed the guy that you are replacing. If you try to flee, I will take care of you, too."

The only way out of this slavery is to escape, a risky venture. Escapees often drown, like the Asian fisherman discovered in the South Atlantic, and recaptured fishermen are severely beaten and, in some cases, killed.

Escaping while the boat is at port is difficult because brokers regularly reapprehend and return the men to the boat. Fishermen who successfully escape might report their abuse to the anti-trafficking police officials, but in many cases, even after fishermen clearly identify the name and number of a fishing boat, the authorities are powerless to do anything because the owners regularly register their boats under duplicate numbers.

I felt that the best way to learn about conditions at sea was for me to travel to Cambodia to meet the former captives in person. Hammond told me about an NGO in Southeast Asia that protects Cambodian workers from exploitation. The NGO officials agreed to talk with me on two conditions: I must not reveal the name of the organization or the identities of the former slaves. I agreed and quickly set a date for my trip.

My flight landed in Siem Reap, Cambodia, and I headed over to the hotel. Surrounded by palm trees and ponds, the hotel grounds were scenic, but the air was sticky and the heat was suffocating. Still, after settling in, I walked around the town to see the real Cambodia. The avenues were crammed with traffic; cars and motorbikes scampered across dirt roads lined with one-room shops offering basics like food, clothing, and everyday wares. I recognized human warmth in the eyes of the people, but I could see that the country's long history of violence and atrocity had left its mark. The story of Cambodia is a tortured one, and its most horrific part occurred under the watch of a man named Pol Pot, a revolutionary-turned-politician. He led a group for more than three decades, from 1963 to 1997, called the Khmer Rouge, the communist party that controlled the country. Pol Pot wanted to create an agrarian socialist society and did so in the most vicious way possible, forcibly relocating Cambodia's urban population to the country to work on collective farms. Those who were opposed to his new government were killed. A combination of malnutrition, beatings, and horrendous working conditions led to the deaths of 3 million people, or 40 percent of the country's population. At the end of his life, Pol Pot supposedly said, "I was responsible for everything so I accept responsibility and blame but show me, comrade, one document proving that I was personally responsible for the deaths."

One of Pol Pot's legacies is agriculture, the mainstay of the Cambodian economy, which accounts for 90 percent of the country's gross domestic product. Rice is a key commodity. In addition to feeding the country, rice is also Cambodia's most important export. Over the last two decades, Cambodia, which is now headed by the Cambodian People's Party, has made strides toward diversifying; however, most rural households depend on agriculture and related subsectors to survive. The average per capita income is only $1,400 per year, and Cambodia ranks as one of the worst places in the world for organized labor.[1] At the same time, the country is a victim of other countries' insatiable appetite for cheap labor.

Later that afternoon, I met my contact at the NGO in the open-air lobby of my hotel. A modest breeze joined us, and the hotel's sari-clad staff served us tea and food. My contact, whom I will call Sam, was in his early thirties. He had jet-black hair and carried himself with a calm, almost insouciant demeanor that belied the urgency of his work. I told Sam that I was an eager student, and under the swirling blades of the hotel's fans, he explained how trafficking works. Traffickers prey on the vulnerable, the unsuspecting men most in need of help, he told me. Traffickers tell them about employment opportunities at home and abroad— legitimate-sounding jobs in construction, agriculture, real estate, and other industries. These words are like honey on a razor blade, promises of well-paying employment that end in predictable tragedy. Typically, these young men sneak across the Cambodian border into Thailand without proper papers, which puts them at the mercy of the teams that traffic them. Victims end up serving on fishing boats against their will.

Each man is assigned to a boat and, if necessary, taken by force. Their fate is not unlike that of the galley slaves of the Venetian empire: "No pay and backbreaking work," Sam explained. Beatings lie in wait for them daily, and they could end up in perpetual servitude. Without enough sleep and food, their daily existence is brutish.

Here is a typical story of a Cambodian tricked into working on a fishing boat. The man is seventeen years old, a boy really, who has never left his village. A trafficker tells this young man about a great job waiting for him in another country like Thailand. The young man is convinced that he will now finally be able to get a job, but he has no transportation. The young man might ask the trafficker, "Will you bring me into the country?" The trafficker says, "Yes, I can help you, but it'll cost you $500." The young man doesn't have that money, so the trafficker says, "We'll work it out later." The victim then asks, "Okay, what am I going to do in Thailand?" The trafficker makes up an appealing answer and the young man agrees. He is told to go to a particular location at a particular time. He waits at the designated spot. A truck arrives to pick him and others up. He is told to hide in the back of the truck, and his long trip begins.

Hours or days later, the truck arrives at a port. The human trafficker pulls back the tarp and says, "Get out, and go quickly into that building." The victim is locked in a room in the building, but it gets worse. The migrant is Cambodian, and the traffickers are Thai.

Because they don't speak the same language, the young man has no idea what's going on. He's stuck in a room with a bunch of strangers. Captives who try to escape are pursued, beaten, and in some cases killed by gang members. The traffickers maintain close relationships with local authorities, who have captured and returned escapees to their captors.

The next thing the captives know, they're being marched up onto a ship. They think to themselves, "Wait, this doesn't look like a construction site," but it's too late. The boat heads out to sea.

Burmese, Laotians, and Cambodians receive the same treatment. Upon arrival at the destination, migrant workers—unable to speak Thai, unsure of their surroundings, and fearful of arrest by Thai authorities or violence by the traffickers—are virtually helpless and accept whatever assignment is offered to them. A number of interviewed fishermen reported that they didn't know they would be working on fishing boats until the trafficker delivered them to the pier. In the "travel now, pay later" system, trafficked men will have to work off the fees they have generated to get onto the fishing boats. In some instances, multiple traffickers are involved in the trafficking of a single individual, each serving as a link in a chain of actors delivering the migrant from the border to the destination.

The high demand for labor on fishing boats means that traffickers are able to sell migrants to captains or other boat representatives for 10,000 to 30,000 baht per person, or an average of US$600 per worker. Research suggests that written contracts are not used by Thai fishing boat owners and that all employment is on the basis of verbal agreements between the boat owners or captains and the fishermen. In the absence of a written contract, fishermen are effectively excluded from social welfare provisions, such as social security. Most verbal agreements oblige a fisherman to stay on the boat for eighteen, twenty-four, or thirty months, with payment made in a lump sum at the end of the agreed time frame.

Pay is either calculated at a standard rate per month—usually between 11,000 and 12,500 baht (US$350–450)—or as a percentage of the value of the total catch after the boat's costs have been deducted, including the cost of food and any medical expenses.[2] However, many times the fishermen are not paid at all, because they owe various hidden fees. For instance, fishermen out at sea for longer periods of time take cash advances from boat captains for living expenses; these advances are then deducted from the total agreed-upon salary.

Once the trafficked men are on the boats, the real horror begins. Life at sea is horrendous, and working conditions are long and arduous. Fishermen are expected to work eighteen to twenty hours per day, seven days per week. Only when the nets are down and recently caught fish have been sorted are the fishermen allowed to eat or sleep. Living quarters on fishing boats are cramped; rooms are divided into squares for the crew, providing only enough room for a small hammock strung from the rafters or a narrow bunk on the cabin floor. No toilets exist on small or medium-size fishing boats; the only facility is found in a small cabinet, which covers a hole in the floor. The need to conserve fresh water and food on long trips means that both hygiene and nutrition are poor.

Several fishermen reported a trip to Indonesia during which they weren't allowed to sleep for a period of three days. Case studies from seventeen Cambodian fishermen identified as victims of trafficking who were helped by another NGO to return from Malaysia confirmed that the men were allowed only two to four hours per day for sleeping and eating. Sheer exhaustion and dangerous working conditions on boats lead to injuries. According to the fishermen, when they fell sick or were injured, little or no medicine was available beyond basic items like pain reliever, and the captain would not stop operations to seek medical treatment for them. Most captains did not allow fishermen to take breaks, and those who were unable to work were often abused verbally and physically by the captain.

I was so absorbed by our conversation that I was startled when our waitress interrupted us to ask whether Sam and I wanted to order more to eat or drink. Though I had been taking copious notes during our two-hour conversation, there was a limit to how much I could absorb in one session. We decided to pick up our conversation later. Sam had set up meetings the next morning with a few men in the countryside. Before going to bed, I sat on the second floor, watching the sunset. The sky was a fiery red.

The next morning, I grabbed a quick breakfast in the hotel. It was already hot at 7 a.m., so I grabbed a bunch of water bottles. Sam appeared with a colleague and a driver. We drove west through the countryside, the same rural areas where many Cambodians are tricked into becoming fishermen on vessels. As Sam explained, "The Cambodian man would like to earn some income to support his family since there are so few paying jobs. He is an easy prey for labor recruiters."

We arrived in a small village to meet fishermen who were held as slaves. I stayed on the beaten path since the country has land mines left over from previous conflicts. The main dirt road of the shantytown was lined with shacks, which were really nothing more than one-floor-high slats of wood thrown together. There was no electricity or running water. In front of the corrugated-roof shacks, on porches covered with cloth, mothers lounged with their children.

Sam introduced me to six men who were held as slaves on boats. One man was held for one year and eight months. Another was held for six years, while a third man was held captive for close to eight years. A common link among these men is the violence they had to endure regularly. I spoke with one man who had thick black hair and was clad in a white shirt. Though he was probably in his late thirties, he looked like a man in his fifties. He described his beatings matter-of-factly, as if he were offering his testimony for the hundredth time. He pulled up his pant leg to reveal a deep round scar. "I was stabbed right here," he said, pointing to the inside of his left thigh. This was no ordinary scar. The man's skin had been torn away. The scar was 3 inches wide and about 2 inches deep.

By this time, it was about noon. I took another swig of water, finishing my third bottle of the morning. My shirt was drenched with perspiration. Another seaman sat down and shared his experience with me. "I did not get the salary throughout the year while I was working on the boat," he told me, the anger welling up inside of him. As he told his story, his voice continued to rise. "And when I asked them to give me payment, they said, 'We haven't calculated your payment yet.'" He was never paid at any point during his year at sea. And yet his experience was less ghastly than what had happened to others.

As I spoke to another man, he shaded his eyes with his hand, squinting against the sunlight. He told me he had a scar on his head, the result of getting struck with a pipe. He pulled back his hair from his forehead, revealing a 4-inch-long horizontal scar below his hairline. The injury made it difficult for him to see in bright light, he said. I shuddered to think about the other damage the blow did to his brain.

In broken English, Sam translated interview after interview relating these common experiences. One man said, "Once we are on the boat, we are beaten and forced to use drugs in order to work long hours." The man with the scar on his head said, "They beat me. I would like to sleep, but they beat us so that we would stay awake at night. You can see my scar. They forced us to use drugs, in order for us to stay awake to work. If we can't work, they will shoot us and throw us into the water."

I asked the NGO worker if it was true that the use of drugs is widespread. "It's widely practiced," she said.

"It's almost work policy, for lack of a better term. The boat owners want to get the most out of their slaves, and what better way to keep them working nonstop than to ply them with drugs, such as amphetamines, just to work longer hours."

The NGO worker, who was on one of the boats, said, "The ship we were on had a lot of boys—some were as young as fourteen—on the boat. They're working all the time, and it's easy to make a mistake that causes a cut. Sometimes it is worse; a rope or line can cut off a hand or other limb. You know, they could get medicine, but the captains only want to give them drugs just to keep them working longer."

Another aspect of life for these seamen is prostitution, which is pervasive on fishing vessels out at sea for a long time. Vessels whose main cargo is prostitutes are euphemistically named "love boats." The NGO worker said, "When the vessels come to port, the crew can't get off the ship. And so these guys have been at sea a year, two years sometimes, no contact with others of any sort. And often [they] can't get into port, so the services, if you will, are brought out to them. And those services will include prostitutes and drugs and booze and food they otherwise wouldn't access, and sort of the whole range of options of things that could be sold."

In the Persian Gulf, for instance, the NGO worker

witnessed, over several hours, a love boat traveling from vessel to vessel. "And there's an unspoken rule, which is, if the captain partakes, then the prostitutes are allowed to come on board."

After our meeting, we drove to meet a young man and his family in a village deeper in the Cambodian countryside. I was drenched in perspiration, so the air-conditioning in the car was a relief. We drove along a narrow road framed by green luxuriant fields of rice. Palm trees reached into the sky under bright white billowing clouds, a scene of remarkable beauty yet a world of great suffering for many. We drove to the house where the man lived with his mother. The two-room house sat on stilts and had a wide wooden ladder that led up to the sleeping quarters. Underneath the house was an open area. Off to the side, a large cooking pot hung on a tripod over burning wood. Nearby a pig, tethered to a stake, grunted at my presence, and a clutch of chickens squawked as they scurried around.

In the seating area under the house, the son and his mother told me their story. The son said that he left his village at nineteen for what he thought was a job in construction. Instead, he ended up serving on a fishing vessel as a slave. He vomited almost nonstop his first few days on the boat, partly because he wasn't used to being at sea and partly because he was processing the shock of his current situation. Sick or not, he was forced to work without a break. Sometimes, he told me, he and the other fishermen were still pulling in lines in the middle of the night and were then expected to start working again only a few hours later, when the sun broke over the horizon. The food was meager, more often than not just scraps of fish, barely enough to keep him alive. If he didn't work fast enough, the captain would punish him. On the boat, the young man was beaten three times, each time more severely than the last. "I always cried when I think of my home, my mother, my family. I cried strongly."

The mother recounted how her son's sudden disappearance affected her. "My son left without telling me any information where he had gone," she said. "I just assumed he had gone to Thailand. He was missing for seven years. I thought he was dead." I asked her how she handled the uncertainty. She said poignantly, "When I was thinking of his death, I would like to die with him. But I could not because I have many children to support and to take care of." I could see in the woman's eyes that, if she could have, she would have laid down her own life to protect her son. Her words struck me deeply and gave me a new admiration for the woman.

The young man told me he thought about trying to

escape almost all the time, but there were never any opportunities. Fortunately, for reasons he doesn't quite understand, his captors finally let him go. Despite his terrible misfortunes, this young man was luckier than most. He lived to tell his tale. Home now after seven years at sea, he is training to be a bike mechanic, while most Cambodian men face limited employment opportunities. "Usually," Sam told me, "they work as migrant labor, pulling carts, or construction worker, or working the plantation. [I] just feel pity for them. I cannot believe such modern slavery still happens on the seas. Most Cambodian people never realize."

As another NGO worker told me, "If you're a poor villager and you just lost one to three years of your life and you went into debt to get those jobs, you have a life-altering problem." I wasn't surprised, then, when the son told me he has gotten into a number of fights at work. Given what this young man has gone through, I suspected he was suffering from post-traumatic stress disorder, though he assured me he is making slow but steady progress.

We left his home and drove back to my hotel. During my final days in Cambodia, I thought over what these men had endured. I kept thinking about one man in particular, who told me his sons were still enslaved on a boat. His anger was palpable, and his voice rose as he spoke. "In my family, there are four of us who went fishing overseas. And only me and one of my sons escaped. And two of them are still missing. And I cannot find where they are." It has been more than two years since he saw his two boys.

I departed Cambodia heavy with sadness and stunned that slavery still exists today. But my contacts at Greenpeace confirmed the accounts I heard. John Hocevar, director of Greenpeace's ocean campaigns, told me that his organization has interviewed dozens of former slaves.

With so many fish stocks in decline, it is now costlier and more difficult for fishing companies to turn a profit. As a result, they have been looking for ways to cut costs, which unfortunately means not paying workers a living wage or providing them with adequate food or water. For some, it means not paying workers at all.

"Slavery at sea is widespread, geographically," Hocevar said. "It's not just Thailand. We've seen problems on Taiwanese vessels. The victims can be from anywhere, really, but especially places like Myanmar, Cambodia, Philippines, Indonesia. We see on Spanish vessels there have been issues, in the South Atlantic, and even the United States is not immune to this problem. We're not exempt from responsibility. It's amazing, but there are more people living in slavery today than at any other time in human history. And there's always been slavery at sea. But part of the reason why we are seeing this resurgence now is environmental, as much as anything."

Greenpeace has interviewed and worked with dozens of former slaves, and their accounts of captivity continued to disturb. "We've talked to people who were thrown in freezers and kept there for so long that, eventually, they lost their fingers," Hocevar said. "We've talked to people who said that they had worked alongside people that were beaten to death and thrown over to feed the sharks. It's not just what happens on the boats. We've also talked to a lot of people who've had similar experiences in processing plants, in some cases. This has been a big problem in Thailand, especially with the shrimp-processing plants, where, you know, one woman tried to escape and she was literally dragged back to the plant by her hair."

I asked Hocevar if men ever tried to escape the fishing boats. "There's just no way to," he said. "Especially when you have a captain or mates that are . . . particularly awful. People will be beaten. They could be denied food. They're going to have to work . . . eighteen-hour days. They won't get a decent place to sleep. You hear about people who are having to eat bait to survive. You know, literally chained to railings of boats and sold from one boat to another."

The most shocking revelation from these interviews is the number of fishermen who have personally witnessed a murder on the boat. "Imagine you interview everyone you've ever worked with and ask, 'So, have you ever seen a murder?' No, of course not. But for these people, that's a remarkably common occurrence."

I flashed back to an interview in Cambodia. A man told me he wanted to leave the boat but couldn't. "I only stayed on the boat. I didn't dare to go anywhere. I saw them kill people, and they are not to get off the boat. I saw the floating dead body on the sea."

I wondered why the captain would kill his own crew. "The men want to stand up for themselves, and at that point, that's where the violence comes in," Hocevar explained. "The murders are to keep the crew—we call them slaves—in line, and partly it's to send a message to the rest of them. If you speak up, if you complain, if you fight back, this is what's going to happen to you."

The connection between slavery and sharks is clear: money. While many countries have outlawed selling and buying shark fins, a lucrative—and insatiable— black market still exists. Crew members get paid next to nothing, often as little as $150 or $200 a month, which doesn't amount to a living wage. Aboard the ship, then, the crew operates under an implicit—often explicit—understanding that they are allowed to sell the fins, splitting the proceeds from the sale among themselves.

Hocevar mentioned a fisherman he interviewed in the Philippines. He said the sideline clause in his contract allowed him to sell shark fins. Those fin sales bumped his monthly wage up by a hundred dollars. After deducting his fees, the shark fins accounted for the only money he earned. Desperate men carve up millions of sharks to collect a few breadcrumbs for their labor. Slaves commit grisly acts of finning and killing sharks every day. What is the result? The ocean loses one of its most important predators, and men lose years of their lives with nothing to show for it but a couple hundred bucks in their pockets and a body full of scars.

The root problem is that poor and politically unstable countries feed vulnerable men to the world's industrial fishing machine. Whether it's in Myanmar, Cambodia, or any other country, men are looking for a better life. Poverty forces them into indentured servitude, and in desperate need of income, they are compelled to poach shark fins, which ultimately hurts the environment.

Tuna, men, and sharks occupy three sides of a triangle. To generate profits, tuna suppliers use slaves to fulfill the demand for cheap tuna products around the world, including the United States. Because the slaves need to supplement their income, they fin sharks and sell them on the black market.

The one party that benefits is the American consumer. Tuna is cheap, and it's cheap for a reason. The labor behind the fishing is free. Yet if one adds up the real cost in death and destruction, that tuna is very expensive. Blood is figuratively all over the can.

The fishermen don't fare as well as the American consumer. These young men suffer grave labor abuses ranging from wage theft to beatings to death. The villagers who were recruited onto these boats go off to sea for years at a time. When they come back without having been paid, they suffer from a crime that should be taken seriously.

When I was in Cambodia, I asked Sam what it was like for these men when they returned home after so many years. Sam said that when they return home, it's hard for them to adjust after having been at sea for so long. Also their health condition is not good, as they have been damaged physically and psychologically.

Another loser in this blue triangle is the tuna, which is suffering from overfishing. The world has launched an armada of vessels that keep getting larger. Bigger vessels mean high operating costs and bigger capital outlays. To offset the higher capital costs, the owners need to catch ever-increasing amounts of tuna. Every day, thousands of vessels catch millions of tons of tuna, dwindling the ocean's stock.

The sharks are the biggest losers, though, the casualty of the shark-fin soup market, which in turn imperils the health of the world's oceans.

The US State Department publishes an annual report entitled *Trafficking in Persons* (TIP), which examines the $150 billion illicit trade around the world. Put together with the help of NGOs, advocates, and survivors of trafficking, the report discusses atrocious conditions in the sex, agriculture, mining, and fishing industries. The 2018 report devoted a section to people trafficked to work in the fishing industry. The caption under a photo of a man in a cage read: "Thai and Burmese fishermen are detained behind bars in the compound of a fishing company in Benjina, Indonesia. The imprisoned men were considered slaves who might run away. They said they lived on a few bites of rice and curry a day and were confined to a space barely big enough to lie down until the next trawler came and the traffickers forced them back to sea."[3]

Governments around the world know that slavery at sea is rampant. Thousands of fishing vessels plow through the waves on their killing mission almost twenty-four hours a day. These rusted, fetid vessels overflow with rats, fish guts, and the dank stink of human sweat and feces. The stench of decay and death follows these boats mile after mile through the high seas.

Society's lack of compassion for our fellow people and for animals leads to an ocean where sharks are dying by the millions and slaves are being abused by the thousands. The connection is basic: a disregard for life itself. The maelstrom of death that descends on the sharks is inevitable when society disregards the lives of workers on the high seas. If you have no respect for human life, then how can you have respect for the lives of sharks?

Chapter 10
Water, Water Everywhere and Not a Tiger in Sight

The Great Barrier Reef is one of the seven wonders of the natural world, the largest underwater ecosystem in the world, and the only living thing on earth visible from space. At 1,400 miles long, it comprises over 3,750 individual reef systems and coral cays and hundreds of picturesque tropical islands with perfect beaches. Because of its natural beauty, the Great Barrier Reef has become one of the world's most desirable dive destinations, which makes it the best place to study tiger sharks, the reef's apex predator and, as such, its fiercest natural custodian.

My friend Rich Korhammer moved from New York to Australia. I called him up to see if he wanted a visitor. Rich told me it was the perfect time for me to come to Brisbane. Located on the east coast of the country, Brisbane is the capital of the Australian state of Queensland and the third most populous city in Australia, with a population of 2.3 million.

When I arrived in Brisbane and saw its palm trees, deep blue skies, and white sand beaches, I liked it right away. The ocean is a translucent turquoise, and with hot summers and moderately warm winters, people can swim and dive year-round. Even though I was several thousand miles from home, I felt I was still in America. Pizza joints, shopping malls, and traffic jams dot the landscape and evoke the Jersey shore. As soon as I settled in, Rich and I headed out to Byron Bay, one of the most famous beaches in eastern Australia. I was jet-lagged and too wired to sleep, so I decided I needed a restorative swim. Three miles wide, the beach at Bryon Bay forms a semicircle of rolling surf, white sand, and, nearest to the road, luxuriant green trees, which offer welcome shade for bathers. Because the surf was rough, my friend and I decided to paddle out on a kayak, but the waiting waves quickly made our 14-foot Sun Dolphin irrelevant. Every time we tried to paddle out, a wave pushed us back, establishing an unbreachable wall of resistance between us and the open water. Undeterred, my friend and I beat on until a 10-foot-high swell forced us to abandon the kayak, mid-paddle, in the surf. Bobbing in the white water, secure only in our life preservers, my friend and I watched forlornly as our vessel-turned-flotsam hurtled back toward the shore, 100 yards away. And yet as I saw the kayak tossing and turning, it dawned on me that, if Rich and I had stayed in the boat, we could have been seriously hurt. The boat could have struck either of us on the head. While it was embarrassing to be forced to swim back in to retrieve it, at least Rich and I were still afloat, relatively unscathed, save for a slight bruising of our egos. The ocean, I was reminded as I swam back to shore, can be a dangerous place, and fear is ever-present. In my case, fear saved us from a potentially disastrous outcome, and I was soon to learn that fear, nature's secret weapon, plays a very important role in the ocean, as well.

My next visit was to Heron Island in the reef's southern stretch. From Brisbane, I drove 340 miles north along the east coast to the small town of Glad-stone, where I caught a ferry for the two-hour ride to the middle of the Coral Sea. Named after the birds that inhabit the island, Heron Island is a circular bay covered in green. Thanks to its status as a World Heritage–Listed Marine National Park, the island is protected. Its flora and fauna live freely, in the wild, largely unmolested by human hands or endeavors, even as scores of scuba divers visit the island, whose wide variety of dive sites makes it one of Australia's most attractive diving destinations. An amazing array of animal life, including green turtles and loggerhead turtles, is abundant throughout the year. Birds like the noddy tern and shearwater muttonbird dart noisily in and out of the island's lush *Pisonia* forests.

As I walked along North Beach, I watched with amazement as small turtles made a run to the water. By sheer luck, I had arrived during turtle-hatching season. Female loggerhead turtles deposit eggs in the sand, and after two months, the hatched turtles start making their way to the ocean, unsupervised and solely on instinct. Before these hatchlings can reach the water, however, they first have to make it through a gauntlet of birds in search of an easy meal.

Standing on the shore, I watched a baby turtle, about three inches long, frantically trying to get to the water. I marveled at how nature can produce a perfect miniature replica of a 300-pound adult turtle. As it scampered to the ocean, its flippers flailed like some frantic rower trying to get to shore. To my horror, a seagull picked it off. After seeing a few other seagulls snatch other baby turtles, I decided to escort a few. I knew I was interfering with the Darwinian struggle of nature, but the turtles, outmatched, needed a hand. I guided one turtle into the gentle waters lapping the white sand and quietly celebrated its christening.

Later, as I walked around the wharf, I caught sight of a fin with a white tip at the top, a whitetip reef shark on the prowl. Nearby swam an Australian garfish, a slender, foot-long fish that always reminds me of the rulers my classmates and I used in elementary school. The whitetip spotted the garfish and surged through the water after it. The garfish made a run for the shallow pools of the nearby coral, skimming inches below the surface as the whitetip tried in vain to catch it. The garfish disappeared into an opening, and at the last second, the whitetip swerved to avoid ramming into the coral. The shark had lost the hunt, as often happens to the predator, but I knew it would soon find other prey.

As I watched the whitetip swim away, I heard the voices of children. A ferry had deposited a class of schoolchildren around ten or eleven years old. The children were headed for a swim off the wharf. Concerned, I ran over to their teacher. "You can't let them swim," I warned her. "There are sharks right along the pier." She looked at me and without a thought said something I never expected. "Oh, that's alright," and she turned away to continue to talk to the children.

"No, you don't understand. There are sharks right

here, and the children are going to jump right in the middle of them."

She remained kind but firm. "Okay, yes, I know. It's alright." I thought she was going to pat me on the head like one of her students.

The children dropped their towels and robes and jumped into the water, making a huge splash. Another whitetip swam 15 yards away, an easy striking distance. But instead of stalking the children, the shark turned suddenly and headed out to sea, more annoyed with the Aussie children's frolicking than hungry. One by one, the sharks in the area quietly swam away, disappearing into the horizon.

Early the next morning, I met several others at the dock to dive Heron Bommie, the island's signature site, which Jacques Cousteau called one of the top ten dive sites in the world. Tiger sharks live there, and I was excited about the prospect of diving with one. In addition, plenty of other species abound, including hammerheads and whitetip and blacktip reef sharks. The site is famous for its coral gardens and underwater pinnacles that start only 15 feet below the surface. With a large oxygen tank strapped on my back, I flipped backward over the gunwale of our boat and headed down the anchor chain. No sooner had I cleared my nose from the pressure change than I saw a school of blue-green damselfish skimming over the reef. From one angle they were a light blue; from another, green, and so vibrant that they glowed. A hawksbill turtle appeared. The sunlight danced on its dappled brown-and-white carapace.

I continued to flow along the seafloor. Various wrasses—cigar-shaped fish wrapped in rings of vibrant blue, yellow, and purple—popped in and out of the coral. A yellow-and-blue-striped cleaner wrasse courageously dove into the gills of a large brown grouper to feed on the parasites and other growths found between the gill slits. I had arrived at a major cleaning station for the fish in the area. Cleaning stations are spots on reefs where pelagic, long-distance travelers are greeted by creatures that clean them up, not unlike a roadside car wash. The 6-inch-long wrasses, which are some of the most colorful fish in the ocean, have specialized mouths that project out and grab, pick, and eat the barnacles, parasites, and other growths that have accumulated on animals crisscrossing the oceans. The fish have been clocked working three- to four-hour shifts, servicing about 2,000 clients a day.[1] Ordinarily, such bright colors on a fish would attract unwanted attention, but in this case, the color is a sign to the dirty, weary travelers that they've arrived at the right location. This symbiotic relationship gets the pelagic fish cleaned up and ready for their next trip, while the wrasses get all they can eat. A Pacific manta ray, the largest of all rays, undulated by, its wings a giant black tapestry with white fringed tassels. Late for its cleaning appointment, it lazily floated by, unconcerned with its tardiness. Several cleaner wrasses approached it, and the ray stopped flapping its wings, like a freighter thrown into neutral, coming to a full stop. The wrasses commenced picking.

I stopped to admire one of the reefs, which was covered with majestic white staghorn coral, whose cylindrical branches can grow to 6 or 7 feet long. It was like looking at a wall of deer antlers mounted on wooden plaques, which I'm sure will help explain how the coral got its name. Inside a small crevice, I came face-to-face with a damselfish. At 5 inches, a damselfish isn't as exciting as a shark, but the specimen in front of me more than made up for any disappointment with its radiant beauty. Tiny dots of luminescence purpled its navy-blue skin, and it ruled this rock like a master of its universe, feeding on zooplankton, every inch a royal. Ignoring the size differential between us, it lunged at me. I flinched, and it won.

Continuing my dive, I floated over a giant clam, the world's largest species of mollusk, measuring in at 4 feet long and more than 500 pounds. Its soft muscle tissue, sandwiched between two hard shells, was a vibrant purple. I had been warned not to stick my foot into its gaping shell, for it can close quickly.

No sharks appeared, but various schools of brilliantly colored surgeonfish swirled around the reef. They graze on the algae growing on the rocks, an important step in maintaining the health of the reef. Each species of surgeon features unique coloring. Some, like the palette surgeonfish, possess a brilliant blue with swirls of equally vibrant navy blue on their sides, like Dory from *Finding Nemo*. Others, like the unicorn fish, are trimmed with a screaming orange color along their dorsal. Whatever the color, all surgeons share a common attribute: a scalpel-like spine running horizontally up the tail, a powerful weapon always at the ready. This blade, which can run as long as 2 inches, is dangerously sharp. One quick thrust can slice open an enemy fish or a diver. Fortunately, I continued through the reef intact.

Swimming between the soaring walls of coral, about 60 feet below the surface, I felt as if I were walking down Fifth Avenue in the long shadows of towering office buildings. Majestic, breathing, and alive, coral is composed of tiny polyps that have a central mouth surrounded by a set of tentacles that grab food from the water. Each polyp is tiny, only a few millimeters in diameter and a few centimeters in length. The supporting skeleton of living creatures can be internal, as in humans, or external, like that of cicadas. In the case of polyps, their skeleton is external, and they use calcium carbonate from the ocean to build up their bodies. Like schoolchildren, tiny coral polyps are growing every day. But at the slow rate of 0.2 to 1 inch every year, polyps need thousands of years to grow into a coral reef.

In that moment, I was content to admire the beauty of the coral—and the beautiful work in progress—as I drifted along the different avenues. While looking to my right, I turned left down one boulevard, and like a tourist bumping into a local, encountered a 10-foot whitetip that had been hiding out in an underwater cave. I moved out of its area as quickly as possible, hoping the shark somehow hadn't seen me. I quickly realized, however, that my clumsiness had captured its attention. The shark swam out of the cave and turned in my direction. I thought it was coming after me. But it swam in a circle and, as it finished its second lap, alighted slowly on the sand. I realized that I had woken it from its nap, and it was more interested in resting than in taking a bite out of its underwater interloper. Scientists do not know if sharks sleep, but their activity varies. Whitetips are generally more active at night and use the day for resting. At 10 feet long, it could have inflicted a lesson for violating napping etiquette. But instead, it just decided to ignore this primate from above and return to its nap. I nearly collided with a shark, and I am none the worse for wear.

Eventually, I joined the group and slowly started my ascent back to the surface, making sure to exhale to rid my lungs of the oxygen tank's compressed air. The sunlight turned the group's air bubbles into a string of blue pearls, which dangled invitingly from the surface. Sadly, my dive had come to an end, yet I left the water exhilarated. I saw the reef's beauty, a natural wonder few things on land can match, a majestic underwater cathedral that needs to be preserved and protected at all costs, lest it falls into ruin.

All life depends on organisms that harness energy from the sun to produce food. They are the key link that captures the energy of the sun and turns it into flesh and blood. The foundation of the sea's food chain is phytoplankton, free-floating single-celled plants and bacteria that capture the sun's energy and, through photosynthesis, convert nutrients and carbon dioxide into organic compounds. Though largely invisible, phytoplankton account for almost 90 percent of the oceans' biomass.

The predominant forms of phytoplankton are algae

and diatoms. There are more than 20,000 species of diatoms. Because they are made of silica, they are transparent and are often called "algae in glass houses." Phytoplankton—literally the grasses of the sea—are the primary producers of the organic carbon that all animals in the ocean food web need to survive, and they produce more than half of the oxygen that we breathe.

One level up the marine food chain are zooplankton, the animals that feast on the sea's abundant phytoplankton. Copepods, microscopic zooplankton that scientists hypothesize to be the most abundant single animal species on earth, are the primary consumers of diatoms and larger phytoplankton. Both phytoplankton and zooplankton are tiny and form the base of the entire food pyramid. Above this level, things get interesting because these various groups of animals are visible to the naked eye. Above zooplankton on the food chain are small carnivores—sardines, herring, menhaden, barracuda, groupers, and snappers. The simple fact of ocean life is that the big fish eat the little fish. Other large consumer groups like green turtles and manatees feed on the seagrass that grows along the coastline. Other herbivores include surgeonfish and parrotfish, which feast on algae on the coral.

The large predators sitting atop the ocean food chain are the "3F group": creatures with fins (sharks, tuna, dolphins), feathers (pelicans, penguins), and flippers (seals, walruses). While these apex predators are large and have unique skills for catching prey, they account for only approximately 1 percent of the ocean's biomass. Understanding how reefs like this function and the role that sharks play is important. Reefs have the highest biodiversity of any ecosystem on the planet—even more than tropical rain forests. Twenty-five percent of marine life lives around coral reefs, which occupy less than 1 percent of the ocean floor. Coral reefs are important in other ways: they provide millions of people with food, protection from storms, and revenue from tourism. An estimated 6 million fishermen in ninety-nine reef countries and territories worldwide—over a quarter of the world's small-scale fishermen—harvest from coral reefs.[2]

The process at work in coral reefs is best described as "emergence," a concept discussed by philosophers and biologists alike since the time of Aristotle. Emergence explains how structures and entities self-organize into complex systems. These structures—reefs or rain forests, for example—are biological systems that, despite being composed of individual species, function as a cohesive whole. The advantage is that the members not only reap the benefits of the stability that the system brings but also rise collectively to successfully meet threats to the system. Emergent behavior arises because of simple and spontaneous interactions among constituent parts. It starts with the ability of individual components to work together within a larger system. Even without an overall controller, these systems can function as a cohesive unit. Bird flocks, ant colonies, and beehives are examples of emergent systems. The reef is a perfect example of an emergent system because it thrives through an interconnected web of creatures that work individually to maintain its structure at all four levels of marine life. It grows and rejuvenates on its own, without the aid of a specific individual commanding the operation. The concept of emergence explains how reefs have survived over millions of years. The system as a whole trumps its component parts, though sharks nevertheless play a very important role. In fact, sharks are the keystone species in the emergent system, as they sit at the top as an apex predator.

In spite of its resilience, the reef is vulnerable. Greg Skomal, the marine biologist we met in chapter 1, explained why during one of our conversations: "In any healthy, natural ecosystem, you get a balance of components that have evolved over eons, to be in perfect balance. And that includes the lowest levels of the food chain up to the highest. You . . . may be able to get away with removing some of the middle sections, because something will fill in, but you remove those ends, and you're going to get a collapse from either end. So you take away those top predators—the sharks, typically—and you're going to get what we call top-down impacts on that ecosystem. It will disrupt the entire equilibrium of the system."

A healthy reef system is a well-orchestrated system that operates magnificently on its own. Spotted and striped fish alike dart in and around coral, working in unison to maintain the reef. Herbivores like parrotfish and surgeonfish employ their sharp beaks to scrape away algae from the coral. If algae is left unchecked, the reef turns ghostly, its colors drained away with only the telltale red pockmarks of ruin left behind. The brightly colored parrotfish, which look as if they are clad in a clown's costume, make another significant contribution to the marine ecosystem. They scrape at the reef with their beaks to remove algae, bacteria, and pieces of rock. They break the rock down into tiny pieces, then swallow and digest these scrapings before excreting the sediment as sand. As parrotfish move around the reef, they stream swirling clouds of sand, which tumbles to the seafloor. Over the course of a year, a single parrotfish can excrete 1 ton of sand,[3] and as thousands of parrotfish excrete thousands of pounds of sand, islands start to form. Parrotfish account for as much as 70 percent of the sand on some islands in the Pacific Ocean.[4] When a sandy beach is formed, it is literally a beachhead for life.

What happens when sharks are removed from the reef? In Australia, when the shark population declined because of overfishing, scientists recognized a disturbing, and somewhat confusing, trend: the reef's herbivore populations dropped in tandem with the disappearance of the sharks. Logically, fewer sharks should have allowed for greater number of herbivores, but the opposite happened.

The loss of sharks resulted in an increase in the number of mid-level (or meso-) predators, most notably snappers and groupers. In the absence of sharks, their natural predator, snapper and grouper populations spiked, which endangered the parrotfish population and, through the concept of emergence, the stability of the entire coral reef.

Without the parrotfish to keep the reef clean, the fleshy algae establish themselves, and disease-causing microorganisms gain a foothold, which exacerbates the problem. The microorganisms alone will kill the coral.[5] In other words, more algae equals more microbes, which equates to more dead coral. Once the coral is dead, fleshy algae are free to expand, colonizing even more patches, and a dangerous negative feedback loop begins. Moreover, the algae and the microbial community associated with it will use up all the organic carbon, robbing the reef of this essential element.

Studies have shown that coral reef ecosystems with high numbers of apex predators tend to have greater biodiversity and higher densities of individual species. By preventing one species from monopolizing a limited resource, predators increase the species diversity of the ecosystem. A healthy coral system requires a healthy number of sharks, which in turn balance the herbivores and the fish that prey on them. Healthy shark populations aid in the recovery of coral reefs whose futures are threatened throughout the globe.[6]

There are economic benefits to a healthy reef system, too. The Hawaiian coral reefs generate significant dollars through tourism and commercial fishing, and of course they improve the biodiversity of the marine ecosystem. The annual benefit of the reefs to the Hawaiian economy, according to one study, equaled $385 million.[7] What's more, the authors of the study assessed the value of the reefs over a fifty-year time period at a whopping $10 billion.

If the Great Barrier Reef—valued by Deloitte Access Economics at $56 billion in Australian dollars—were to devolve into an algal reef, it would likely die completely over time, a staggering ecological and economical loss. "Without the sharks, you collapse the basic foundation of the entire coral reef ecosystem," Skomal told me. "This simple example shows how removing the top predators will cause problems for the ecosystem, and possibly kill it."

When sharks are healthy and present, the emergent system that governs the reef and the seagrass allows the marine ecosystem to function normally. However, when sharks are removed, the inevitable trophic cascade occurs, in which change flows from the top of the food chain all the way down to the bottom. And with serious gaps in our understanding of the oceans, these disturbances can lead to disastrous consequences. We could be proceeding apace with the destruction of sharks without full knowledge of the impact of their demise, only to recognize the changes to the food web when it is too late.

We might want to rethink our beliefs about apex predators and their role in the ecosystem. The antiquated view that predators are killers and that we are better off without them has been proved wrong. Apex predators don't just kill; they also give life to the ecosystems where they rule. All the creatures that live in an emergent system are governed by a delicate interconnection. The ecosystem desperately needs sharks, and the inexorable question becomes, Do we humans need them as well to ensure our own survival?

Just as the removal of sharks has catastrophic consequences for reef systems, the same is true on land when apex predators like wolves are eliminated. Perhaps the most revealing story of the importance of apex predators took place in Yellowstone National Park when wolves were decimated at the turn of the twentieth century. It is unfortunate that the park officials did not have a chess player's ability to think ahead. Yellowstone officials could not imagine the havoc their actions would cause when they viewed the wolf as an enemy that needed to be exterminated. The eradication program was brutal. Wolves were trapped, shot, and poisoned. By 1926, not one wolf remained in Yellowstone. Even though bears and cougars also preyed on the park's elk, the absence of wolves took the foot off the brake on the elk population. Once the wolves were gone, the elk population exploded, and they devoured young shoots, bushes, and trees across the landscape.

As attitudes toward wild ecosystems changed, people began questioning whether a wolfless environment was a healthy one. Biologists in Yellowstone began exploring the idea of introducing Canadian wolves to the park, and on January 12, 1995, the first eight wolves arrived from Jasper National Park in Alberta, Canada. The impact was immediate. Scientists reported that a trophic cascade occurred in Yellowstone with the return of the wolf. The resulting change was dramatic, and scientists are still trying to understand it.

The first impact was that the elk population was brought under control. The quick recovery of plants like berry-producing shrubs and animals like beaver and lynx soon followed. Bare, dusty ground was replaced with luxuriant aspen, willow, and cottonwood trees. And the songbirds and migratory birds returned, and the sky was filled with their songs.

Various animal species on the verge of collapse sprang back to life. Some of the effects were indirect: when wolves abandon their prey after feeding, scavengers such as coyotes, bald eagles, golden eagles, grizzly bears, black bears, ravens, magpies, and red foxes finish off the carcass, and this food source is essential for survival in winter.[8] Mountain lions and grizzly bears do not leave much behind when they are done with their prey.

An exciting and relatively new scientific discovery is that apex predators do not have to make a kill to affect the ecosystem. One of their functions is to change animal behavior, which can have a powerful effect on molding the environment. That change is driven by one of the most powerful emotions in the animal world: fear. Without the wolves in Yellowstone, the elk didn't have to move around when browsing in winter and could eat as much as they wanted of the young willows and other plant species. By the time the elk were done, the beavers were not left with much to feed on to survive the winter, and the beaver population suffered.

When apex predators are nearby, the animals at the lower trophic levels have to be wary. Death is constantly stalking them. When the wolves were reintroduced in 1995, the elk needed to be cautious; they couldn't browse on plants at their leisure and eat them into the ground. As predatory pressure from wolves changed elk behavior and kept them on the move, the willows flourished, and so did the beavers. The wolves did not have to kill the elk to achieve this result; the change in elk behavior drove the reestablishment of the beavers. And the impact of the wolves continued to cascade down, right into the water and the rivers.[9] To the astonishment of scientists, the rivers meandered less, and erosion was reduced. More trees were able to survive along the riverbank, and their roots kept it in place. The deer avoided the sunlit valleys for fear of being trapped by wolves, and vegetation could now take root. These plants stopped even more erosion. With a stronger riverbank, the water flowed straighter.

Adult wolves, which weigh less than 100 pounds, are not the largest predators in Yellowstone; however, they are giants when it comes to moving entire rivers and breathing life into the ecosystem.

The lessons we learned about the wolves in Yellowstone apply as well to the ocean. The herbivores and carnivores that live on the reef are upended with the loss of sharks, just as the beavers, lynx, and bears in Yellowstone were when the wolves disappeared. Are there other locations around the world that further prove this link? It turns out that there are. On the other side of the world, another Yellowstone exists under the sea. I was soon to find out that fear works just as well underwater as it does on land.

A young scientist named Mike Heithaus, PhD, is studying the role of sharks in Western Australia from his home base at Florida International University. Specifically, he's looking at the impact of tiger sharks on Australia's seagrass ecosystem. Pictures of tiger sharks and notebooks filled with twenty years' worth of data line his office. His greatest research project, which took place in Australia, took ten years to complete. At least twice a year, he travels halfway across the world to study seagrass in a unique marine environment. Centuries ago, before humans started affecting the environment, seagrass flourished along coastlines across the planet.

Traditionally, seagrass is viewed within the scientific community as something of a poor stepchild. Few scientists research it, primarily because seagrass isn't as evocative, or inviting, as coral reefs and mangroves, which offer scientists opportunities to travel to exotic locations like the Maldives. While seagrass is found along coastlines all over the world, excluding Antarctica, I have yet to hear of scuba excursions off the Belt Parkway to explore the seagrass in New York City's Jamaica Bay. Still, seagrass offers interested scientists like Heithaus exciting new discoveries about its importance to the environment.

"First and foremost, [seagrass] provides food and habitat for small fish, shrimps, crabs," Heithaus told me when I asked about the relationship between sharks and seagrass. "They grow up to be important in food webs but are also species that people want to catch and eat. And so we need to make sure we're protecting seagrass ecosystems for that reason."

If one is going to study seagrass, Shark Bay is one of the best places to do it. Designated as a UNESCO World Heritage Site, Shark Bay has one of the largest seagrass beds in the world, teeming with dugongs, dolphins, sea turtles, and of course tiger sharks. Located

500 miles north of Perth on Australia's western coast, it's not for tourists. The russet-colored land is literally deserted. Visitors are faced with two options: study seagrass or die of boredom, slowly.

Heithaus and his team have been working at Shark Bay for years. On his most recent field trip, Heithaus examined the relationship among tigers and their prey, most notably sea turtles and dugongs, or sea cows, the biggest grazers of seagrass. Under a bright-blue morning sky, he jumped into a 17-foot-long aluminum skiff along with his three young assistants. He gunned the engine, and the wind blew back his light-brown hair and split the whiskers of his goatee. From my position on the skiff, Heithaus, a former All-American swimmer, looked very much at home at the helm. He steered the skiff into the calm bay to begin the day's work. The first order of business was to set the hooks to catch the tigers, which required special and controversial equipment: a dozen drum lines. Only drum lines are strong enough to hold a shark in place. This line consists of a floating drum (a barrel) filled with a rigid polyurethane foam, which keeps it buoyant. This barrel has two lines attached to it. One is attached to an anchor on the seafloor, while the other features a large baited shark hook. Six-inch-long steel hooks, which are attached to a steel chain, are baited with 2 pounds of fish to attract sharks. Watching Heithaus and his team set the line, I thought back to my childhood when my father taught me to fish in Cape Cod: slice off a 1-inch piece of a squid, attach it to a small hook, and pray. If he and I caught a 3-pound, 12-inch scup, we celebrated a hand-clasping victory. Heithaus's prize, in comparison, was a 1,500-pound catch.

Once Heithaus and his team set a dozen lines, they got to work setting up the cameras. After checking for tiger sharks, the professor jumped into the shallows, planting cameras in waterproof boxes in the seagrass. The cameras would allow him to monitor how the sharks interact with the fish, turtles, and dugongs.

The next day, the crew hopped back into the boat. As the boat approached the drum line, Heithaus pulled back on the throttle. The drum line was bobbing and moving ever so slightly. Something was underneath the float. The scientist cut the engine and, as the boat drifted in the bay, grabbed the float, revealing the tiger shark's telltale black stripes against its gray skin. The tiger, a female, was nearly as big as the boat, but this didn't deter Heithaus. "Look at that! A good, beefy shark," he said gleefully. He maneuvered the boat alongside the shark and, with his bare hand, flipped her over and put her into tonic immobility, a kind of trance sharks fall into when the electric pores on their snouts are rubbed and their stomachs are exposed. To make sure the tiger was not injured, Heithaus restarted the engine and slowly dragged the shark through the water, pouring life-giving oxygenated seawater over her gills.

With the shark safely secured to the side of the boat, the entire team got back to work. Heithaus latched an acoustic tag onto the shark's dorsal fin. If the shark is caught again in the future, the tag will let them know how much it has grown. It will also let Heithaus and his team track the shark's location in the open sea. A junior graduate assistant measured the shark: 10 feet long. With any luck, the shark will grow to 17 feet, matching the record-setting length for a tiger. A second assistant worked the shark's tail. She punctured the shark's skin with a needle. "I got the blood," she yelled. Their work completed, Heithaus pulled out a 4-foot-long steel cutter. "Keep your hands away from the mouth," he warned his assistants before snapping the steel linked chain in the shark's mouth. Freed, the shark rolled over and, still groggy, slowly swam away, no worse for the wear.

Heithaus motored the boat away and continued the process of checking the drum hooks. Before lunch, the team tagged seven tigers averaging 10 feet. In total during his study, Heithaus has tagged 700 tiger sharks and never lost even a finger.

Over the next few months, the videocams and the tracking data provided a rare window into the animals living in the seagrass. All this information had something to say, and Heithaus shared with me Shark Bay's fascinating secret.

Sea turtles and dugongs, which look like nature stuck a vacuum cleaner on their nose, are found in warm coastal waters from the western Pacific Ocean to the eastern coast of Africa. They feed in Shark Bay on the rich nutrients in the seagrass. However, dugongs can easily overgraze on seagrass.

"In those areas where the sharks aren't as dangerous and the big grazers spend their time, you have very little seagrass," Heithaus told me. "It looks like a really closely cropped lawn—not a lot of carbon being buried, and also, not a lot of place for the little fish to hide and grow up."

He continued, "What we found is that the tiger sharks change the behavior of where these big grazers feed. Where there are lots of tiger sharks, it's really dangerous for the grazers. When the sharks are around, those grazers spend almost no time there, and that protects the seagrass. So we get these really big, dense forests of seagrass that provide great habitat and also sequester lots of carbon dioxide."

As proof of the impact, Heithaus put out cages to keep the grazers out. Those areas were turned into this lush salad bowl of seagrass. "What that showed us is that the tiger sharks really are controlling not just where the big grazers are spending their time, but actually, the seagrass beds themselves. So, what we're seeing in this one situation is that sharks are probably critical to maintaining the health of oceans."

I asked Heithaus what would happen if tiger sharks died off. Something else would surely step into their place and pick up where they left off, right? "Well, in some places, it looks like if you were to lose sharks, other animals could step in and fill that void," he said. "So, out in open-ocean ecosystems, if you lose pelagic sharks, to some extent, billfish and tuna could increase and fill a similar role. We're not sure that'll happen, but it's certainly possible.

"When it comes to big animals like tiger sharks, however, there is really no other animal that can fill their role. There's no other species out there that can threaten adult sea cows. There's no other species that can tear through a turtle shell so effectively and control their populations. Not threaten them by eating them, but actually scare them. Because it's not just about the body count or how many prey sharks eat; it's about how they change the behavior of the prey and where they're feeding. So, for an animal like a tiger shark, there really is no replacement."

After I left my interview with Heithaus, I played the skeptic in my mind. Anyone who's ever tried to lay down a towel on seagrass hates the brown clumps that wash up on the beach following a storm. Is seagrass really that important?

Even though seagrasses and seaweeds look superficially similar, they are very different organisms. Seagrasses belong to a group of plants called "monocotyledons," which include grasses, lilies, and palms. Like their relatives, seagrasses contain leaves, roots, and veins, which transport nutrients and water throughout the plant and contain little air pockets that keep the leaves buoyant.

Sharks were there to witness the first seagrass that evolved 100 million years ago when some flowering plants moved from land into the ocean. Even though these plants were under the water, they still produced flowers, pollen, and seeds. Seagrass is usually found in shallow water of less than 10 feet; however, some species can thrive in dim light at depths of up to 190 feet. Over tens of millions of years, seventy-two types of seagrass, from small to large-bodied seagrass, evolved, allowing them to spread across the planet. The seagrass meadows then altered the seafloor, building up soil and supporting entirely new ecosystems.

Just as the sand from the parrotfish allowed islands to form and generate more life, the seagrass allowed ocean life to take another step forward by acting as a nursery habitat. Schools of fish flourished off the nutrients tucked in the seagrass meadows. Sea turtles could grow fat off the long, flat, ribbonlike blades. Animals like manatees took up residence. Seahorses latched onto the paddle-shaped leaves. Juvenile fish hid from view in the spaghetti blades and grew up and reproduced. Shrimp and invertebrates clung to branching shoots to live and breed, and they ate the algae that grew on the seagrass. The green blades could then take advantage of the sun's rays to grow, creating a positive feedback loop; life flourished throughout the habitat, the ocean's third most valuable habitat behind estuaries and wetlands. A single acre of seagrass can support 40,000 fish and millions of tiny invertebrates.

Seagrass is a valuable weapon against climate change. The seagrass pulls carbon out of the atmosphere and uses that carbon to build up more seagrass, just as trees use carbon to grow trunks. Even when seagrass dies, it gets buried in the seabed, which traps carbon in the ocean rather than the atmosphere. Each year, the world's seagrass meadows can capture up to 83 million metric tons of carbon, and a single acre of seagrass can sequester 740 pounds of carbon per year, the same amount emitted by a car traveling about 3,860 miles. A terrestrial forest, depending on the tree species and age, captures anywhere from 500 to 2,500 pounds of carbon per acre per year, as a comparison.[10] Seagrass sequestration may be at the lower end of the terrestrial range but it still makes a significant contribution in the fight against global warming. While seagrasses occupy only 0.2 percent of the total ocean floor, they are estimated to be responsible for up to 10 percent of the organic carbon buried in the ocean.[11]

Seagrass also plays an important role in maintaining fisheries around the world. The plants draw fertilizer runoff and other pollutants out of the water, locking them safely away in meadow soil. In 2014, a University of Cambridge study on seagrass meadows in southern Australia estimated the value that seagrass provides to the fishing industry at $87,000 per acre per year.

In some polluted areas, the sediments that flow into the ocean are loaded with deadly hydrogen sulfide, a corrosive and flammable chemical compound. By pumping oxygen into the seabed, seagrass helps detoxify the chemical. Scientists estimate that an acre of seagrass provides more than $11,000 worth of filtering every year.[12] On an acre-by-acre basis, seagrass meadows are more productive than today's fertilized cornfields.

Seagrass also helps fight disease on the reef. Like any living organism, reefs are susceptible to pathogens. Seagrass purges from the ocean pathogens that can threaten coral reefs. Recent studies have shown that many sea plants produce natural biocides. In lab studies, the abundance of bacterial pathogens of multiple marine fish and invertebrates in seawater was lower when seagrass was present compared to a paired site without seagrass. The net effect of seagrass is to remove potential pathogens of marine invertebrates that can damage the reef and the fish. When seagrass meadows were present, a 50 percent reduction occurred in the relative abundance of potential bacterial pathogens capable of causing disease in various marine organisms.[13]

Outbreaks of diseases that affect reef-building corals are a considerable driver of global reef degradation. The losses in the Caribbean and Indo-Pacific are approximating 50 to 80 percent. For example, one bacterial pathogen isolated from sewage is linked to the decline of two dominant reef-building corals now on the US endangered species list. Regardless of the mechanism involved, alleviating coral disease is vital for the well-being and livelihoods of the 275 million people living within 20 miles of coral reefs as well as providing a direct benefit to reef-dwelling species.[14]

My trip to Australia finally came to an end, and I headed home to New York. I now have new respect for those brown clumps of seagrass on the beach. With Australia's low human population, seagrass is in relatively good shape there. However, in places like the Indian Ocean, where shark populations are in decline, increasing populations of sea turtles are causing entire seagrass beds to virtually disappear. Between 1959 and 1976, when Hawaii culled nearly 554 tiger sharks, turtles took to eating most of the nutritious seagrass, which led to the destruction of seabed ecosystems. Scientists in Hawaii contended that as the tiger shark populations dwindled, the health and balance of the marine ecosystems was negatively affected.

Other scientific studies in various regions of the world corroborate Heithaus's study in Australia. For instance, models of dietary shifts of harbor seals in response to sleeper sharks provide another example of shark intimidation resulting in behavioral change. The modeling study looked at seals and sleeper sharks in the northern Pacific Ocean. Seals feed on both herring and pollock but due to their larger population prefer the latter. However, to catch the pollock the seals have to dive deep, where they run the risk of encountering sleeper sharks. If seals want to play it safe, they stay at the surface. Scientists concluded that without the threat of sharks, seals would shift to deeper waters, eating more pollock and less herring.[15] This is another example of the removal of sharks causing cascading impacts throughout the food web.[16]

Seagrass and coral reefs are two highly visible areas where sharks play a role. However, sharks have other effects on the marine ecosystem that are not as visible. Sharks prey on the sick and weak members of the local ecosystem. By removing the sick, they prevent the spread of disease to schooling fish, and they strengthen the gene pools of the species. Since only the fittest survive to reproduce, the fish species's gene pool becomes more robust. In addition, sharks keep the oceans clean. Sharks like great whites and tigers eat dead whales and carcasses of other animals. If the sharks were not there to do their job, the rotting remains could spread disease. The direct impact of the shark on the ocean is substantial, but when we consider the indirect impact, the shark's influence is multitudinous.

The shark's influence even goes beyond the ecosystem to include an impact on the economy through ecotourism. Ecotourism offers an interaction with the environment in a way that reveals animals and nature in their undisturbed state. The experience fosters a greater appreciation of natural habitats and sometimes generates new insight about how we humans are interconnected to the environment. Innumerable jobs are created from ecotourism that centers around marine activities, such as shark and reef diving and snorkeling. A healthy environment is the cornerstone for ecotourism.

The evidence is overwhelming that living sharks drive revenues and job creation. I asked Wes Pratt for his views on whether sharks are important in ecotourism. "A good shark is not a dead shark," he said. "In some communities, a shark is worth a lot more alive than it is dead." Sharks provide tourist communities with significant revenues. "A dead shark is probably worth $100 to $200, depending on the size of its fins and the species, whereas estimates of its value to a community in an ecotourism operation, for example, are between $100,000 and $250,000 per shark, when you divide the worth of the ecotourism, the profits over the years or over an annual basis, by the number of sharks that are used in that community."[17]

These estimates might be too conservative. Islands in the Pacific rely on tourism, and sharks are one of the biggest draws, just as countries in Africa benefit from lions as a draw for safari trips. In total, the annual revenues to the Pacific Island nations engaging in shark ecotourism amount to more than $120 million.[18]

Shark tourism generates $42 million annually for Fiji.[19] To protect this revenue stream, the Fijian government put into place special conservation measures. Diving clubs pay the state in exchange for permission to dive in protected areas. Some of that money is paid to villages lying adjacent to the protected area where sharks congregate. Residents get paid for not fishing those waters as a way to protect the sharks. For some villagers, the payment can add up to tens of thousands of dollars a year.[20]

Another nation that has also come to value sharks is the Republic of Palau, an island group located in the western Pacific Ocean, about two and a half hours west of the Philippines. With approximately 340 islands, the nation forms the western chain of Micronesia's Caroline Islands, a postcard-perfect vision of turquoise waters lapping on white-sand beaches stretching out below verdant mountains. Shark tourism accounts for approximately 8 percent of Palau's gross domestic product.[21] The roughly one hundred sharks that inhabit the country's most popular dive sites were each worth $179,000 annually to the local tourism industry. Assuming each shark lives ten years, that would monetize each shark with an approximate lifetime value of about $1.8 million.[22] Because of this, Palau was the first to establish a shark sanctuary, banning all commercial shark fishing in its exclusive economic zone, a 630,000-square-kilometer area equal in size to France. In May 2015, the president of Palau announced a plan to ban all commercial fishing in its waters as soon as contracts with Japan and Taiwan expired. The South Pacific nation is obtaining radar equipment and drones to monitor its waters.[23]

Other nations around the world run shark-diving operations. A recent study found that shark tourism companies operate in eighty-three locations in twenty-nine countries.[24] The most visible are the whale shark–diving trips, which occur in Africa, Asia, and the Americas.

Shark-based tourism totals $25 million annually in Australia. Even in the United States, sharks generate significant tourist revenue. According to a 2016 independent economic report commissioned by Oceana, shark-related dives in Florida alone generated more than $221 million in annual revenue and supported over 3,700 jobs. New York State, along with the federal government, invested $158 million for a shark exhibit in the New York Aquarium in Brooklyn. Drawing thousands of visitors each month, that exhibit is key to the survival of the entire aquarium. In addition, ecotourists can join a great white shark–diving operation in Montauk, only 90 miles away from New York City. To kill sharks is to potentially damage these job-creating operations.

In summarizing, the total annual economic value of the world's coral reef ecosystems, which include mangroves and seagrasses, is nearly $30 billion, which is composed of several parts. Tourism and recreation owe $9.6 billion to coral reefs; coastal protection, $9 billion; fisheries, $5.7 billion; and biodiversity, $5.5 billion.[25] Damage to these ecosystems would create a major economic loss to humankind, not to mention the psychic loss of such extraordinary habitats. The sharks, as the ocean's guardians, in essence protect our economic interests.

During my research, I revisited the work of conservationist Aldo Leopold. Though Leopold is no longer a household name, he influenced the modern environmental movement. After graduating from Yale college in 1908, he joined the US Forest Service (USFS), where he worked to increase the deer population in New Mexico. Because the Forest Service believed that the best way to increase the deer population was to create a predator-free ecosystem, Leopold and other USFS employees eliminated the local wolf population. They shot all the wolves and he said, "We reached the old wolf in time to watch a fierce green fire dying in her eyes. I re- alized then, and have known ever since, that there was something new to me in those eyes—something known only to her and to the mountain. I was young then . . . and I thought that because fewer wolves meant more deer, that no wolves would mean hunters' paradise."[26]

The wolves' removal had an immediate negative ecological impact, which caused Aldo to change his approach to conservation. Later, as a professor at the University of Wisconsin, he became the nation's fore- most expert on wildlife management, introducing the revolutionary concept of the trophic cascade. His 1949 environmentalist book, *A Sand County Almanac*, was an early call for conservation. Leopold said, "I now suspect that just as a deer herd lives in mortal fear of its wolves, so does a mountain live in mortal fear of its deer. And perhaps with better cause, for while a buck pulled down by wolves can be replaced in two or three years, a range pulled down by too many deer may fail of replacement in as many decades."[27]

The lessons Leopold learned about wolves apply di- rectly to sharks because they play a key role in protect- ing marine ecosystems like the seagrass beds. If sharks are removed, the intermediate predators can multiply and consume the smaller animals that keep the sea- grass blades clear of algae. Cod were overfished in the

Baltic Sea, which allowed the smaller fish to consume the invertebrates that cleaned the seagrass. The seagrass went into decline there. When people think of sharks, their first reaction is fear. Intrinsically, fear is emotionally neutral; its positive or negative influence depends on one's reaction to it. As I learned on my first day in Australia, fear can either serve you well or make you act irrationally in ways that hurt you and others. If Leopold had studied sharks, he might have said, "Just as sea life lives in mortal fear of its sharks, so does an ocean live in mortal fear of losing its sharks." In other words, society's first fear about sharks should be losing them, not being harmed by them. Society should use its fear of losing sharks to make sure they survive to protect the reefs and the seagrasses around the world. If we lose that fear, and sharks are eliminated from the ocean, we'll experience a worldwide trophic cascade that will lead to disastrous ecological consequences. When a shark's dorsal fin breaks the surface, we should breathe a sigh of relief that the waters are healthy and will remain healthy as long as sharks are left alone to perform their role.

Chapter 11
The High Seas

Older than trees, the sharks have populated the planet longer than almost every other living organism. Throughout their 450-million-year existence, sharks have survived five extinction-level events, but today, the species is staring down its greatest threat.

In the past fifteen years alone, the scalloped hammerhead, great white, and thresher shark populations are believed to have fallen by more than 75 percent. During this same period, the oceanic whitetip shark population has declined by at least 80 percent. All told, according to Boris Worm, professor of marine conservation and biology at Dalhousie University, the number of sharks killed per year is 100 million, though it's possible the number could be as high as 273 million.[1] The continued massacre of sharks around the world each

year is the result of several factors, but the biggest culprit is the overexploitation of the oceans.

Far from the eyes of the world, fishing vessels prowl
the high seas, which is defined as ocean areas outside
the 200-mile territorial jurisdiction of coastal nations.
The high seas officially cover 64 percent of the world's
ocean surface. Across the planet, one-third of all marine fish stocks are fished at unsustainable levels, and
the remaining two-thirds are maximally sustainably
fished, meaning there is no more fish that can be removed without harming the yield.[2] This practice is
proving ruinous to local economies and the world's fish
stock. It is also putting the shark species at the greatest
risk at any time in its history.

China plays an outsize role in this destruction. As
one of the world's largest suppliers of seafood, including seafood destined for the United States, China also
imports and reexports the vast majority of the world's
shark fins. Controlling 20 percent of the world's seafood market, China showcases an armada of 200,000
fishing vessels and 2,500 distant-water vessels. Ten
times the size of the United States' distant-water fleet,
China's fleet is unrivaled on the high seas and, argu-
ably, in the history of humankind.

Despite possessing some of the best fishing grounds
in the world, the United States controls only 6 percent

of the world's seafood market. And, while China's percentage share increases annually, the share of the United States has been in regular decline.[3] Over the previous two decades, China has been responsible for most of the growth in worldwide fish availability per capita. One of the drivers of the dramatic expansion in China's fish production is the country's voracious appetite for fish, a taste that has been fueled by growing domestic income and wealth. Fish consumption in China has increased steadily, rising from 14 kilograms per capita in 1993 to 40 kilograms in 2013.[4] That's an average annual growth rate of 5 percent.[5]

Global Fishing Watch, a nonprofit organization that monitors global fishing efforts, spent five years monitoring fishing ships from various countries. They estimated that Chinese vessels spent 17 million hours fishing in international waters in 2016, which easily exceeds the nine other leading countries combined. In comparison, Taiwan, which registered the second highest hours according to Global Fishing Watch, clocked an estimated 2.2 million hours, while the United States, clocking in slightly fewer than 2 million hours, comes in ninth in the Top 15 Fishing Nations.[6] Between 2000 and 2012, according to the European Parliament, China's distant-water fleet caught 4.6 million tons per year globally, or more than ten times the 368,000 tons per year China reported to the Food and Agriculture Organization (FAO) of the United Nations.[7] Chinese fishing fleets account for nearly 40 percent (34.7 percent) of the world's total haul. Combined, the next five leading countries—Taiwan, South Korea, Spain, Japan, and Ecuador—account for 36.4 percent of the world's total. (The United States accounts for 2.1 percent.)[8]

The Chinese distant-water fleet, which roams the world from Central and South America across the Pacific to Oceania to South Korea, puts the world's fish stock at risk. China's fishing bureau, which is part of its department of agriculture, issues licenses to local firms to fish overseas. With its huge fishing fleet, China has an outsize impact on various fisheries around the world, especially in areas that lack local competition. Chinese vessels have been well documented overfishing in other counties' exclusive economic zones, or EEZs. Africa is a prime example. The ocean waters off West Africa represent one of the most productive fishing grounds in the world. The cold water from the Benguela Current mixes with warm water from the South Equatorial Current. That mixing of oxygen-rich cold water with warm water creates a surge in sea life. For centuries, African artisanal fishermen have launched their small brightly colored boats to fish these waters, taking only enough from the blue waters to feed their families and communities. That world has been upended. An unfamiliar new silhouette has appeared on the horizon, and a new sound of throbbing engines is now heard above the surf. An armada of Chinese fishing vessels now dominates the waters of West Africa with more than 500 industrial fishing vessels operating in the region. Their ships have swooped down and taken huge catches with estimates as high as 2.8 million tons worth $7 billion.[9] Local fishermen are squeezed out of their livelihoods along the entire coast of Africa by the might of the Chinese. Impoverished countries like Ghana and Guinea-Bissau are unable to find enough fish to satisfy the demand for food and jobs. Yet Africa is not alone in suffering at the hands of China's fishing operation.

But what happens when those firms violate local fishing laws? Usually nothing, unless the Chinese are forced to respond to a particularly blatant violation of the law. Indonesia, for instance, is trying to crack down on illegal fishing in its waters. In the past five years, Indonesian officials have seized and destroyed approximately 500 fishing boats. While several countries such as Vietnam and Malaysia are fishing illegally in their waters, Chinese vessels have also been implicated. Recently, Indonesian authorities seized the 598-ton fishing boat *Fu Yuan Yu 831*, which had 35 tons of fish and protected tiger sharks on board. The Chinese fishing boat had flown the flag of Timor-Leste and had on board five other national flags—those of China, Indonesia, Malaysia, the Philippines, and Singapore—an action that violates international law. The boat is listed as being owned by Fuzhou Hongdong Pelagic Fishery Co. Ltd., which is based in Fuzhou, China.[10]

While the consumption of shark-fin soup has fallen in China—although consumption is, admittedly, difficult to quantify—due to increased conservation and an official ban of shark-fin soup at state functions, this decline is offset by the rising demand for the soup in markets such as Singapore, Hong Kong, and Thailand. Wealthy urban Thais are now regularly serving the soup at business meetings and family events, and in Macau, the country's booming casino and hotel industry has led to a spike in demand from wealthy patrons for the soup. (Of the thirty new hotels built in Macau, twenty-eight feature shark-fin soup in their restaurants.)[11]

At the same time, sharks are now regularly targeted as a primary catch—and not just for their fins. To keep up with the international demand for seafood, fisheries have been going after shark for its meat. Between 2000 and 2011, for instance, trade in shark meat rose an astounding 42 percent.[12] Brazil, in particular, has experienced an eightfold rise in shark meat imports since 2000, making it the world's largest importer of shark meat.[13] Over this same period, Spain emerged as the world's third largest producer of shark catch, behind only Indonesia and India. From 2002 to 2011, Spanish fleets captured more than 61,000 tons of "sharks, rays, and chimeras," exporting a larger percentage of its shark meat to European markets, most notably Italy, the continent's top consumer of shark meat.[14] While China consumes the most shark fins, and imports and reexports the vast majority of shark fins around the world, the country was only the seventh largest importer of shark meats from 2000 to 2011, though, according to the Food and Agriculture Organization of the United Nations, its shark meat export industry has been trending upward in recent years.[15]

In addition to their fins and meat, sharks are killed for squalene, which is used to manufacture cosmetics like lipstick and lotions, and for their skins, which are used to produce everything from shoes and handbags to sandpaper.

Caught in the vise of globalization, sharks and shark-derived products are shipped around the world like fruits and vegetables. They are now a big business, and the total value of shark exports today is worth approximately half a billion dollars.[16] Even factoring in a marginal decrease in the consumption of shark-fin soup around the world, then, the parallel increase in demand for shark meat and other shark-derived products means the casualty rate of sharks around the world remains stubbornly high.

If this current rate continues, some shark species could face further degradation. The pelagic, or deepwater, sharks are particularly vulnerable. Of the thirty-nine species of pelagic sharks, seventeen are threatened with extinction, largely due to fishing on the high seas.[17] One of these is the thresher, which is currently listed as vulnerable on the International Union for Conservation of Nature's Red List. Further exploitation of this shark can result in humankind losing a remarkable predator, a marvel of the animal kingdom. The thresher attacks its prey with an elongated tail, whipping its sicklelike tail over its head at a remarkable 80 miles per hour. The thresher is the only ocean creature with the ability to kill with its tail. If the thresher were to disappear from the world's oceans, its absence as an apex predator would cause an almost immediate trophic cascade. The loss of this magnificent species could prove disastrous to the world's oceans—to say nothing of what its loss would mean psychologically and emotionally to humans, who are unable or unwilling to act as responsible stewards of the planet.

To prevent such immeasurable losses, how fishing is regulated on the high seas needs to change—not just to protect sharks—but commercial species like tuna, too. As currently constituted, the management of fishing advances the short-term interests of certain states rather than the long-term sustainability of fish stocks around the world. Protesting voices are ignored, and it's akin to killing a guard dog to keep the sheep safe.

The regulatory framework is made up of a patchwork of entities called regional fisheries management organizations (RFMOs), which were established when we believed ocean resources were unlimited. These seventeen international groups are made up of countries that share a financial interest in managing and conserving fish stocks. Of these groups, five are the so-called tuna RFMOs, which cover 91 percent of the world's oceans and manage fisheries for tuna. They are also responsible for sharks, seabirds, and turtles affected by fishing.

One of the largest RFMOs in the world, the Western and Central Pacific Fisheries Commission (WCPFC) covers the western and central Pacific, where most of the world's tuna fishing takes place. According to the National Oceanic and Atmospheric Administration, bluefish tuna is overfished in the Pacific. The annual monetary value of the tuna caught in this region is

$6 billion. Driven more by decisions to protect vested interests than preserving the long-term viability of the tuna industry, RFMOs like the WCPFC have failed to reduce and restrict the number of fishing vessels at sea. Moreover, they ignore the scientific advice to limit catches, because heeding such warnings would constrain fish-processing industries. The RFMOs have failed to prevent overexploitation of migratory fish stocks, to rebuild overexploited fish, and to prevent the further degradation of marine ecosystems.

Shark management in the Atlantic is no better than in the Pacific. The International Commission for the Conservation of Atlantic Tunas (ICCAT) governs all commercial fisheries in the Atlantic Ocean and Mediterranean Sea for the group's fifty-two contracting parties, the affected nations, and other governmental organizations. Like the Pacific's RFMOs, the Atlantic commission is responsible for the conservation of tuna and sharks and other maritime species. All the major shark species, including great whites, swim in these waters. And, while scientists cannot estimate the great white population in the Mediterranean, great white sightings are occasionally made around the Greek islands, the Strait of Sicily, and the Strait of Gibraltar. Critically threatened, the white shark is close to being ranked as endangered.

Unfortunately, data used to determine the rate of fishing mortality, a key parameter to gauge the health of shark stocks, is self-reported and, therefore, often unreliable. Fishermen may underreport or even not report shark captures. A study was undertaken to figure out the mortality rate of the mako population.[18] Tracking data of forty satellite-tagged makos showed that they travel so widely that they swam through the management zones of nineteen different countries, which illustrates the importance of countries working closely together to conserve makos. More important, the data showed that 30 percent of the forty satellite-tagged mako sharks[19] were killed in commercial fisheries, ten times the previous estimate. The death rate for any mako over a calendar year in the North Atlantic is 30 percent.[20]

Recognizing this, the World Wide Fund for Nature (WWF), which also advocates for the protection of the oceans and more sustainable maritime activities, calls for a complete halt of all mako catches in the North Atlantic to help increase the chances of rebuilding the stock by 2040.[21] This recommendation is in line with the same advice of ICCAT's own scientific committee. In addition, WWF is calling on ICCAT to adopt further precautionary measures, such as the implementation of a long-term management plan to protect threatened blue sharks, a species in high demand from the fin soup market. As a natural scavenger, the blue shark helps keep the ocean clean. But because it instinctually goes after baited hooks from longline vessels, this evolutionary task puts the species at particular risk. Every year, 20 million blue sharks are caught around the world. Throughout the Mediterranean, Spanish fishing fleets, which harvest these waters for swordfish, catch blue sharks as bycatch. They then export most of the blue shark fins to Hong Kong, where they can comprise 60 percent of the shark fins for sale.

As a result, the IUCN ranks blue sharks as near threatened, a rank better than makos and threshers. Since the regulatory authorities refuse to develop a plan to save the blue shark, however, it is likely to suffer the same fate as the endangered whale shark. While scientists continue to grapple with the loss of blue sharks, it's likely the sharks' increased vulnerability would lead to an explosion of squid populations around the world, which will threaten the existence of fry and tuna populations.

Sharks are not the only species being overfished in the Atlantic where Chinese vessels roam. The ICCAT's scientific committee provides tuna population assessments to the tuna fisheries' management team annually. What they've found is disturbing. Several tuna species are exploited beyond sustainability. Commercial fishermen in the Atlantic have been overfishing bigeye tuna populations there by 20 percent over the total allowable catch (TAC) in the past two years, even though the ICCAT adopted a recovery plan for the species in 2015. At current levels, scientists estimate that the chances of the stock's total collapse by 2033 is around 60 percent, worse odds than a coin toss. The same statistic is true for yellowfin tuna, which has been overfished for years. The current catch exceeds the ICCAT's allowable limit by 36 percent. Despite clear advice from its scientific committee, and other conservation groups like the WWF, the ICCAT did not adopt a comprehensive management plan to address the overfishing of tropical tuna during its annual meeting in November 2018.

What's worse, that same year, the ICCAT increased the total allowable catch for bluefin tuna to 36,000 tons, the highest total allowable quota by the international commission. While scientists continue to warn that bluefin tuna stock will decrease significantly at this level, ICCAT's new quota takes effect in 2020. The ongoing overfishing of bluefin as well as tropical tuna combined with a massive illegal trade in bluefin tuna have continued unaddressed for years.

The decision to ignore the scientific advice and to postpone any action to address the overfishing of yel-

316 • WILLIAM McKEEVER

lowfin tuna and of the already-threatened bigeye tuna population is likely to undermine the tenuous recovery of these species. It's highly disappointing to see that there is no serious political will to guarantee the full legality and sustainability of these fisheries. Policy makers as well as consumers should be more aware of the value of tuna as a key natural resource, instead of looking at it as a mere commodity to exploit.

Because RFMOs in the Pacific and Atlantic have been unable or unwilling to curtail the overexploitation of the large fishing fleets of China and other countries, smaller island countries and conservation organizations around the world are taking small, but important, steps to do so. Many Pacific Island countries recognize that current fishing practices could put the tuna fishing industry in peril. Actions by these coastal states, including Palau and Papua New Guinea, and a call for a new approach for managing the Pacific Ocean's tuna fisheries can reverse the situation before it is too late. One of the easiest actions for the coastal states is to introduce legislation to reduce the overcapacity of fishing vessels in territorial fishing waters. Another key action is to ban transshipment immediately, at sea and in port. Transshipment, as discussed in chapter 8, allows fishing vessels to launder fish, skirting a country's exclusive economic zone (EEZ) protocol.

Transferring catch to reefers in the open ocean allows vessels to declare their catch as having been caught on the high seas, free from local jurisdiction and taxes. The Chinese vessels make use of transshipment constantly and record the greatest number of encounters with reefer ships.[22] The concomitant loss of earnings to the states in the Pacific are unknown. The lost fees could be anywhere from hundreds of millions to more than a billion dollars. Surprisingly, no RFMO has prohibited the transfer of catches at sea from longline vessels to reefers or other ships. Bubba Cook is the Western Central Pacific tuna program manager for the World Wide Fund for Nature (WWF). He focuses on the sustainable management of tuna throughout the region. "Transshipment on the high seas should be prohibited unless countries can unequivocally prove it is impracticable for the vessel's operation, which they have either been unwilling or unable to do for years, and subject to unassailable oversight," he told me in a telephone interview, adding that transshipment makes it all too easy for illegal, unregulated, and unreported (IUU) fishing to take place.

Many Pacific Island countries have similarly started calling for key areas in the western and central Pacific Ocean to be closed off to all fishing to protect fish populations until existing sea life can recover in full. In the past two years, more of the ocean has been set aside for protection than during any other period in history. In 2017, for example, the governments of Chile, Palau, and New Caledonia either expanded or created marine parks, committing to protect nearly 3.8 million square kilometers (1.47 million square miles) of ocean. The Chilean government, which is a global leader in ocean protection and conservation, signed into law in 2018 protections for nearly 450,000 square miles of water—an area roughly the size of Texas, California, and West Virginia combined. Split into three regions, the largest is the Rapa Nui Marine Protected Area, where industrial fishing and mining are prohibited, but traditional fishing is allowed. More than 40 percent of Chilean waters have some level of legal protection. And, in 2009, Palau became the first nation in the world to create a shark sanctuary. All types of shark fishing are prohibited within the sanctuary, which covers roughly 230,000 square miles, an area equal to the size of France. Palau's action inspired other countries—the Maldives, Honduras, the Bahamas, and Tokelau—to establish their own shark sanctuaries. Officials in Palau, which has a population of only 20,000 inhabitants scattered across 200 islands, want to see a total ban on shark fishing.

"The ocean needs sharks more than it needs people," Wes Pratt told me. "And I would have larger marine protected areas. There was an Australian study that said that you needed to protect at least 40 percent of the Great Barrier Reef in order to have sustainability. We don't come close to that. We've got a lot of great effort and a lot of great people making marine protected areas, but to protect the oceans and the ecosystem and the coral reefs and the whole—everything in between—we need to fence off more of the ocean and leave sharks to do their long-term slow conservative business."

Human activities have led to a global decline in marine biodiversity of approximately 50 percent, roughly half of what it was fifty years ago. The International Union for Conservation of Nature has called for at least a 30 percent increase in the portion of the ocean that is highly protected to help effectively conserve biodiversity. Less than 3 percent of the oceans are currently protected, a dishearteningly low percentage that includes the planet's two largest reserves—the Ross Sea Region Marine Protected Area off Antarctica and the Papahānaumokuākea Marine National Monument in the northwestern Hawaiian Islands—which were designated marine parks in 2016. The United Nations and countries around the world need to increase this percentage to maintain the ocean's health and to protect the livelihood of fishermen around the world.

Unless additional areas are set aside, endangered fish such as whale sharks, the ocean's gentle giant, will continue to suffer from the fishing onslaught. Whale sharks, which are killed for their skins, fins, and oil-rich livers, are currently listed as endangered. Because whale sharks excrete potassium, phosphorous, nitrogen, and other nutrients essential to the health of the ocean, their traveling from the surface to 3,300 feet below distributes these nutrients throughout their underwater ecosystem. Similarly, when whale sharks die of natural causes, their carcasses sink to the ocean floor, where they continue to excrete nutrients as they decay. An increase in designated marine areas to protect whale sharks would allow the species to continue to play a vital role in maintaining this ecosystem far into the future. Without them, their future—and the future of this ecosystem—remains in doubt.

Marine sanctuaries are often opposed by those with an interest in exploiting the ocean's resources. The fishing industry argues that everything must remain until others can prove that a marine sanctuary will improve fish conservation, rather than expecting the international fishing industry—and countries like China who stand to lose billions of dollars in revenue and their healthy percentage of the market share—to prove their aggressive fishing practices aren't depleting the world's fish population. Regardless, the evidence is clear that marine-protected areas help protect sharks, or at least give the 450-million-year-old species a fighting chance at survival.

Some coastal states in the Pacific want an immediate end to longline fishing in their economic zones. They understand the devastation that the longline operations bring. Globally, it has been estimated that approximately 300,000 sea turtles (comprised of 250,000 loggerheads and 60,000 leatherback sea turtles) are hooked and killed by longlines every year. In addition, 160,000 seabirds are killed annually.

If coastal states in the Pacific end longline fishing in their economic zones, fishermen can turn to other effective methods for catching tuna. Pole-and-line fishing has been practiced for centuries. The method involves attracting a school of tuna to the side of a bait boat by throwing live sardines and anchovies overboard. The tuna enter into a feeding frenzy and are caught on the hooks. They are then hauled out of the water, one at a

322 • WILLIAM McKEEVER

time, using pole and line. The size of the tuna caught this way is usually small, mostly consisting of albacore and skipjack, but oftentimes valuable yellowfin tuna are also snared. More important, bycatch of unwanted animals is reduced. One benefit of switching from longline to pole-and-line fishing is the comparatively high employment that would ensue from replacing large-scale industrial fisheries with smaller boats and more manpower. Consumers who want to buy sustainably produced tuna might vote with their pocketbook by buying pole-caught tuna. So coastal states can pursue pole-and-line fishing; however, it cannot meet the world's demand for tuna. In 1950 this method met the demand of 2.5 billion people on the planet, but now with the world's population at 7.7 billion and rising, industrial-scale fishing is required.

As problematic as it is, the worldwide fishing industry is crucial for human sustenance and jobs. Fish provide 3.2 billion people with almost 20 percent of their average per capita intake of animal protein.[23] In addition, 540 million people around the world live by fishing. Countries should no longer be allowed to fish in a way that puts fish species at risk when the technology exists for sustainable fisheries. The community of nations can no longer stand on the sidelines when the behavior of a few nations threatens the world's fishing industries. Why should the livelihood of millions of people be put at risk for the benefit of a few?

If the world wants to preserve the shark and tuna populations, which in turn preserve the planet's maritime ecosystem, a large portion of the world's food supply, and a vital sector of the world's workforce, then we must start action to reform the current inadequate fishing management structure. The RFMOs could easily fix the shark-finning massacre. WWF and other NGOs have proposed that all RFMOs should require vessels catching sharks to keep the shark's fins attached to the body to make sure that meat is not wasted when the ship comes back to port. Technically, the RFMOs hang on to a similar, but ultimately ineffective rule. If a vessel keeps shark fins, they are required to retain enough of the shark carcass to ensure the fin-to-carcass ratio does not exceed 5 percent. The reality is that this regulation is window dressing. The fins are not even required to be stored in the same place as the carcasses. So a pile of dried fins might be stored in the bow of the ship while a mass of frozen shark carcasses is held in the freezer that might only represent a fraction of the representative fins. Because RFMO officers have to count every fin and compare it to every carcass to comply with this ratio requirement, enforcement is impractical and cost prohibitive. If a more comprehensive "fins attached" rule were put into place—working in conjunction with the kind of ban on transshipment Bubba Cook at the WWF calls for—the practice of shark-finning would then become cost prohibitive.

Humans are now the dominant species on earth, and whether we like it or not, this position requires stewardship of the planet. The world is struggling today with illegal, unregulated, and unreported fishing. We can no longer allow China—or any other country—to decimate the shark and tuna species. The planet will not survive if a powerful few countries, including other Asian nations, continue to profit from the destruction of maritime species essential to the health of the oceans. The time has come for the less powerful to rally together to protect sharks, tuna, and the fishermen who legally and sustainably harvest them. The current patchwork of independent RFMOs and individual fishery commissions doesn't encourage the kind of coordinated, comprehensive, and regulatory approach to fishery management and maritime conservation required to meet this increasingly dire emergency. A new approach, modeled on the efforts by the governments of Chile, Palau, and New Caledonia and the work of the IUCN, is necessary to protect the ocean's resources, and in the long run, such a unified worldwide approach will be good for all nations. The time has come to change the way tuna and sharks are managed and put an end to the exploitation taking place on the high seas before it is too late. If we act now, these two valuable species, which belong to us all, can be preserved for the benefit of generations to come.

Chapter 12
Shark Warriors

South Africa displays some of the most remarkable animal diversity on the planet. The vistas of
the Kalahari Desert and Kruger National Park are unlike any in the world. Beyond these remarkable sights
on land lies an equally spectacular environment in the
oceans. The warm current from the east mixes with
the cold current on the western side of the country, and
the mixture of nutrient-rich cold water and warm water
produces a cornucopia of food. Apex predators gather
near the southern coast for its vast array of fish species, including sardines, who travel up the east coast of
South Africa between May and July during the annual
sardine run, the largest biomass trip in the world. The
sardines form shoals, tightly packed swarms of billions
of fish, which attract predators like the shark.

I wanted to go to South Africa to meet with conservationists, scientists, and others who would share their insights into sharks. At least, that's what I told my editor. What I really wanted to do in South Africa was swim with sharks. To learn about these animals in their natural habitat, I mapped out a thousand-mile-long "Shark Route" from Cape Town, South Africa, to Imhambane, Mozambique, along a varied and contradictory coastline of modern golf courses and modern towns, ancient forests, and stretches of deserted white beaches.

My flight from New York was a grueling nineteen hours with a two-hour stopover in London. When I finally exited the plane in Cape Town, a large sign greeted me: WELCOME TO THE MOTHER CITY. To the north is the dramatic flat-top Table Mountain, which dominates the capital's landscape like an altar. To the south are the turquoise waters of the Atlantic, which extend beyond beaches of white sand into the distant horizon.

I immediately visited the capital's famous Two Oceans Aquarium, where I had previously scheduled a supervised dive with ragged-tooth sharks, a common species in the waters off South Africa. They are so common, in fact, that locals refer to the sharks with the sing-song nickname "raggie." The raggie is a cool gray-bronze on its dorsal side and a much lighter hue on its large underbelly, which camouflages the shark from predators approaching from below. But don't let its nickname fool you. The pointy-nosed raggie, which is found in the Atlantic, Pacific, and Indian Oceans, is a fearsome-looking shark. Its teeth are like needles; rather than cutting up prey like its serrated-teeth cousins, the raggie seizes and holds fish, rays, and squid with its teeth before swallowing them whole. Ovoviviparous, raggies fertilize and hatch eggs internally, and raggie pups live inside the mother shark until they are strong enough to be birthed. In utero, the pups cannibalize unfertilized eggs, which the mother continues to produce, a process known as "oophagy." Only two pups are born, one from each uterus.

My dive master, Scotty, has been diving for forty years. With his completely white beard and sizable paunch, he looked like an underwater Santa Claus, though a St. Nicholas with a thick, often indecipherable Scottish brogue. Before we entered the water, as I donned my scuba equipment, Scotty warned me, again and again, not to underestimate the raggies. While considered to be one of the more passive sharks, raggies have nevertheless been known to attack humans, though only a very small percentage of these attacks are fatal. Concerned, I asked Scotty how aggressive the raggies I was about to encounter might be.

"The raggie makes minimal exertion during the day," Scotty told me, "conserving their energy for their nighttime hunting. He is slow moving and spends the daytime hidden in caves, paying little attention to divers." Scotty then deadpanned, "But don't pull on their tails. You're only asking for trouble."

He and I entered the water and swam off along the rock of the underwater canal, settling along the sandy bottom. I looked back up to the surface and watched the bubbles from my oxygen tank hurry to the surface. Just then, a raggie passed overhead. I felt at last the satisfaction of being in the water with a shark and experiencing it not in a book but in the flesh, in its world. I was totally focused on the shark. With the sunlight obscuring some of the shark's details, I could see little more than its silhouette, which lent the encounter a dreamlike feel. Still, I was able to make out the underslung mouth and thinly pointed teeth. Swimming slowly through the water, the raggie seemed relaxed and unhurried, perfectly at home in this underwater environment. Before my dive, I imagined such an encounter would feature some drama, but the raggie simply ignored me. I grew calmer and calmer, and I had to fight the very real urge to touch the shark's underbelly, lest it entertain second thoughts about my presence near it. Most sharks remain afloat because their liver holds a large quantity of oil, which is lighter in density than the water. Not our friend the raggie, though, which maintains its buoyancy a bit differently. Raggies gulp air and hold it in their stomachs to keep them from rolling over or sinking to the bottom of the ocean.[1] Because I was enjoying the dive, marveling at the wonder of the raggie above me, I wasn't paying attention to what was ahead of me. When I finally looked, I saw a shark moving straight at me, a 10-foot-long raggie, with a wide girth, about the size of a sedan. The shark and I were on a collision course. The canal was too narrow for the two of us to pass without bumping into each other. I had to come up with something.

I considered hugging the seafloor to let the shark pass overhead or, as an alternative, turning back the way I had come into the canal. Anything, really, in the hope that the raggie wouldn't take an interest in me. Neither option was ideal, because in such close quarters, the shark was sure to notice me, no matter what I did. Just as I was about to panic, however, the raggie did something unexpected. It turned around.

The shark had been doing the same calculations as I was and realized that we were going to collide. To avoid me, it moved its pectoral fins forward and tilted them

332 · WILLIAM McKEEVER

up, which brought its forward momentum to a dead halt. At the same time, the shark twisted its body to the right, and as it flicked its tail, the shark headed off in the opposite direction. This experience confirmed everything I had learned during my travels. A supposedly deadly predator made way for me? I could have been a light snack, and yet it turned the other way. That's not exactly the man-eating behavior of a *Jaws*-like beast.

My first shark encounter had gone well, but it was time to head out again on the Shark Route.

The next leg of my journey was an hour-long drive south of Cape Town to a town called Fish Hoek, a former whaling and fishing community that hugs the western side of False Bay, a shallow 18-mile-wide basin that forms a horseshoe parabola around the Cape of Good Hope. Today, Fish Hoek is a popular suburb of Cape Town. My drive took me past shopping centers until I finally reached a road along the beach surrounded by hills. I was scheduled to meet Lesley Rochat, a conservationist, activist, and artist who, in 2003, founded AfriOceans Conservation Alliance, a nonprofit that works to protect sharks from exploitation. Since its founding, Rochat has run numerous public-awareness campaigns, using her complete arsenal of skills as a filmmaker and photographer to retell the story of sharks.

By changing people's perceptions of sharks, Rochat believes she can start to save what she calls "the ocean's most misunderstood fish." For her efforts, which range from protests to documentarian action, Rochat has earned the nickname "Shark Warrior."

I first learned about Rochat from a video of her free diving with sharks, which she posted online. Underwater, her brown tresses wafted behind her, like a mermaid's mane, while 8- to 10-foot sharks swarmed her tiny 5-foot frame. I was impressed with her bravado, swimming freely in open water 25 feet below the surface, armed only with a mask and snorkel. But as I was about to learn, Rochat is fearless in most things, matching her underwater actions on land. In another video, a naked Rochat attached herself to a drum hook, a giant hook used to catch large prey. Drum hooks sometimes suffocate sharks because sharks cannot move sufficiently to pass water and oxygen over their gills. The video, which has been viewed on YouTube more than 200,000 times, was part of a national shark awareness campaign to ban the use of drum hooks—"Get Hooked on Conservation, Ban Drumlines"—in South African waters. According to the estimates, drum hooks capture 600 sharks—and other marine life, including dolphins and whales—per year.

I drove my car up a steep hill to get to Rochat's

ranch-style house, which is nestled in the cliff above the beach in Fish Hoek. She greeted me from her porch overlooking False Bay. From this aerie, we gazed out over the blue water toward the cliffs 20 miles across the bay. We walked inside and sat down in her living room. Its wide windows gave us a 180-degree view of the ocean. A former model and television presenter, Rochat has a palpable, contagious energy, especially when she talks about sharks, which fortunately for me was pretty much throughout our hour-long conversation. "I've always had a love for the ocean," she said. "It goes back to when I was a child, and I used to spend hours swimming in the sea. The ocean fascinated me, and I wanted to know what lay beneath its surface. I think if I couldn't live next to the ocean and see it every day like I do, something in me would die."

Of all the ocean creatures Rochat discovered underwater, sharks fascinated her the most—first out of fear. "I was once afraid of sharks," she said, "because I believed the media's portrayal of them. I was so afraid of them that my dive buddies nicknamed me 'Shark Bait.'"[2]

Her initial fear of sharks is understandable though largely unfounded, as I've shown. Like me, Rochat lays much of the blame on Peter Benchley and his novel turned cult movie *Jaws*, which portrays sharks as bloodthirsty killing machines with an insatiable appetite for humans. "Benchley instilled [in people] fear and loathing of sharks, a perception the irresponsible media perpetuate," Rochat wrote on her website.

Long before "fake news" worked its way into the public's psyche, the media played just as pernicious a role as Benchley's *Jaws* in branding sharks as relentless man-eaters. Because sharks are an instant attention grabber, journalist are always quick to report any shark encounter as if it were a real-life *Jaws* attack, no matter how inaccurate or absurd.

If there is any doubt about the media's role in perpetuating the old prejudices against sharks, the media's handling of a series of events during the summer of 2001 serves to remove all doubt. That summer, a bull shark bit off the right arm of an eight-year-old from Pensacola, Florida. A few weeks later, in Volusia County, Florida, twenty-two surfers were bitten while paddling through schools of bait. That same summer, sharks killed a ten-year-old boy in Virginia Beach and a twenty-eight-year-old man in North Carolina. The media swarmed over each incident, and *Time* magazine published a cover story titled "Summer of the Shark." Three years later, after a great white killed a teenager at West Beach near Adelaide, Australia, a *London Times* columnist wrote that the Australian government should kill sharks there. More recently, in 2015, when international surfing champion Mick Fanning was nudged by a shark during a surfing tournament, a TV news reporter in Australia called for the culling of sharks in the vicinity of any surfing tournament.

Meanwhile, producers of film and TV have realized that a gullible public has bought into the myth of the man-eating shark. Like children around a campfire wanting to hear ghost stories, the uninformed are willing to pay to see blood and guts on the big screen. The concept behind *Sharknado*—a made-for-television movie about sharks being picked up in a waterspout and thrown through the air into people's swimming pools and backyards, where they try to eat humans—is of course absurd, but the movie was a big hit. On television, avid shark enthusiasts gobble up any news about shark attacks.

Rochat's once-palpable fear of sharks, however, is just a memory. By plunging into the water, she defeated her fear and got to know the sharks for the gentle, intelligent creatures they are. "They have never stopped fascinating me," she said. To combat the cultural misrepresentation of sharks, Rochat launched her "Rethink the Shark" campaign in 2011, a national advertising push that consisted of a poster series and three television spots produced by advertising giants Saatchi

& Saatchi and Groundglass. One commercial starts off with a familiar, *Jaws*-inspired scene: beachgoers frolicking in the water without a care, until a lifeguard starts blowing a whistle, pointing out a danger lurking in the water. The swimmers panic and desperately try to get out of the water. A crying baby sits abandoned on a beach blanket as running legs flash past her. But then Rochat's commercial turns. Instead of a shark fin, the camera shows a toaster bobbing up and down in the wake. Then the viewer is greeted with a startling statistic: "Last year, 791 people were killed by defective toasters. Nine by sharks." In the two subsequent commercials, the initial scene is repeated, but the statistics reflect fatalities stemming from kites and chairs: 358 and 652, respectively. Each spot ends with the ingenious tagline: "Rethink the Shark."

"Normally, people have these terrible stories about their lives changing because of something traumatic, like losing a limb," Rochat told me. "But mine was something very different. My life changed one day when I met a shark."

Rochat visited Cape Town's Two Oceans Aquarium to take photos for an article she was writing about the plight of sharks. There she encountered a raggie named Maxine with a scar around her gills, the result, Rochat soon learned, of getting caught in the shark

nets of KwaZulu-Natal, a province along the southeastern coast of South Africa. Those nets were established
as a barrier to protect bathers from sharks; however,
rather than steering clear of the nets, sharks would get
entangled in them. Over thirty years, the nets killed
33,000 sharks plus the unintended death of 8,500 rays
and 2,500 dolphins. Though few raggies or gray sharks
survive these nets, Maxine did. A few months later, as
the shark was migrating down the South African coast,
Maxine was caught during an angling competition and
then transported to Two Oceans, where she was kept
in captivity for nine years. Taken with Maxine's remarkable story of resilience, Rochat immediately decided that Maxine would be a fantastic ambassador for
sharks, one that could help raise awareness for the plight
of the species worldwide. Rochat pushed for Maxine's
release, which the aquarium granted, and she designed
a research project around her called the "Maxine Science Education and Awareness Project." Maxine was
satellite tagged, and Rochat followed the raggie on her
underwater sojourn.

"Maxine was the catalyst for me," Rochat said.
"The more I traveled and became aware of what was
happening to the ocean and to sharks, in particular,
the more concerned I became. And I thought, this is
it. I have got to do something about it. Taking photos

and writing about it is not enough. I then packed up my well-paying corporate job and founded AfriOceans, which is dedicated to saving sharks and raising awareness about the plight of sharks all over the world."

It's one thing to produce public-awareness campaigns for sharks, and quite another to strip down to a bikini and swim with sharks, without any protection. I thought of the video of Rochat swimming up to a 10-foot-long tiger shark. I doubt I'd have the courage to go that far myself. I prefer to observe their beauty at a relatively safe distance. Rochat, however, has a remarkable record of swimming safely with sharks—and she has articulated a powerful statement. "I've been diving with sharks for over fifteen years now," she told me. "And I swim with sharks for many reasons. Firstly, I swim with them because I want to show people that the perception they have of sharks is totally wrong. These aren't monster man-eaters looking for people to eat but rather incredibly beautiful animals that deserve our respect, our admiration, and most of all, our protection. Every time I swim with sharks, something special happens to me. It's almost like meditation. It's just the air in my lungs, the water on my skin, and the sharks and me. It's beautiful. The main reason I swim with sharks is, simply, because I love them."

"Isn't it risky?" I asked her.

She answered so nonchalantly that I was sure she has been asked this question thousands of times. She stared right back at me, as if to question how I could even ask such a question. Diplomatically, she answered, "The perception that people have of sharks as monster man-eaters is a global perception. If you go out anywhere, like I do, to speak to people and talk about sharks, the first reaction they have when they see images of me swimming with them and free diving with them is, 'Oh! Aren't you afraid?' or 'Don't they want to eat you?' They just don't understand. And that is why education and awareness is one of the main things that we focus on. Because at the end of the day, if people don't understand and we can't help them to understand, how can we expect them to care and even want to conserve these animals?"

In other words, swimming with sharks in such a vulnerably beautiful way allows Rochat to lead by example, telling her fellow countryfolk, "Here I am with these animals, and I'm coming to no harm. They're not what we think they are. They're incredibly beautiful animals that are very fragile, and they need all the help that they can get to survive."

As an example, Rochat pointed to the sevengill shark, an undercelebrated species compared to South Africa's more famous sharks. "They are just a short swim away from the shore," she said, pointing to the bay below her property. "You can go down, descend to twelve meters, and enjoy this incredible experience with these beautiful animals. They swim to you, and no baiting. They are totally safe." Sevengill sharks are one of the most primitive sharks, with a large, thick body and blunt snout. They live off fish near the ocean bottom. They have seven gills, while other sharks have five.

Recently, Rochat has been making the compelling argument that people will only protect what they love and will love only what they understand. And because people will only understand what they're taught, dedicated education is required. Through her work, she's continuing the cultural rehabilitation and rebranding of sharks from man-eating monsters to vanishing species.

"Sharks are caught in almost every fishery as bycatch, and there is also a targeted pelagic shark fishery in South Africa," she said. "The problem is sharks are not a conservation priority at all, despite the fact that they play a vital role in maintaining the delicate balance of the marine ecosystems, and they're a very good indicator of the health of our oceans. And once our sharks go, we're going to see a repercussion in the entire marine ecosystems. They're still not conserved."

In South Africa, only three shark species are fully protected: the whale shark, the basking shark, and the great white shark. However, compliance, even with those few protected species, is almost nonexistent. "They say there is some management, but it's actually zero—it's on paper," Rochat said. She believes that without compliance the law is toothless. Her view is that this is not just a South African problem; it's a global one. The universal lack of management and compliance reflects the greed, corruption, and ignorance surrounding shark conservation. According to Rochat, the authorities just do not understand why these animals are needed in the marine ecosystems and why we need to conserve them. "That understanding is not there," she said.

Extreme environmental times, Rochat argues, call for extreme measures. In addition to drum lines, Rochat and her AfriOceans Conservation Alliance are targeting longline fishing vessels and the shark nets that continue to line long stretches of the South African coast, the same nets that snagged Maxine and continue to lay waste to the country's shark population. For all their perceived menace, Rochat was quick to remind me, sharks are extremely fragile and in deep trouble. As she pointed out in a TEDx Talk, advanced technology and improved fishing practices ensure that practically no fish, regardless of species, escapes harvesting. There are 233 shark species on IUCN's Red List, twelve of which are critically endangered, mainly because of a lack of fisheries management.

According to Adam Welz, a Cape Town–based journalist, the South African government has encouraged this poorly regulated fishing industry, granting licenses to politically connected operators. Since 2013, Welz reported, three to six demersal (meaning just above the ocean floor) long-liners fashioned with up to 2,000 baited hooks have been working hundreds of miles of the southern coast of South Africa.[3] With so many long-liners operating off the coast, coupled with the country's extensive nets, it's no wonder that shark species like great whites and tigers are disappearing from this once-fecund part of the world. Researchers at South Africa's Stellenbosch University recently found that there are only between 353 and 522 individual great whites left in South African waters. "The numbers in South Africa are extremely low," said Sara Andreotti, a professor in the university's department of botany and zoology and lead author of the study. "If the situation stays the same, South Africa's great white sharks are heading for possible extinction." The study, which was based on six years of fieldwork, is the largest field-based project to date covering the country's great white population.

Like Welz, noted great white expert Chris Fallows blames this decline on longline fisheries. In 2013, the South African government established a program called Operation Phakisa (Hurry Up) to create more jobs, granting six longline fishing permits in 2014, which opened up fishing in areas between Cape Agulhas and East London. These permits led to overfishing of many shark species. To be fair, the South African government did have the foresight to set up a marine protected area (MPA) on the Southern Cape coast, along the De Hoop Nature Reserve west of Mossel Bay. The De Hoop MPA, which stretches three miles into the sea, is critically important to protect marine wildlife. Because visitors come to the dunes to catch a glimpse of the endangered southern right whale, the MPA benefits the country as a tourist attraction. However, the MPA lies adjacent to the areas where longline fishing is permitted. Fishing ships will catch sharks that migrate in and out of the eastern border of the De Hoop MPA. As a result, dive operators have reported a dramatic decrease in great whites and other sharks in the area. The goal of Operation Phakisa may have been to create jobs and jump-start the country's exportation of sharks, but the damage these fisheries inflict to local ecosystems is becoming clear. I remembered my experience with Greenpeace, specifically how the longline fisheries indiscriminately kill all kinds of sea life. The longline vessels are targeting smaller species of sharks like soupfin and smooth-bound species that are a favorite prey of great whites. "These long-liners have increased their efforts in recent years and driven smaller shark species to their demise," explained Fallows. Juvenile great whites, for instance, could be starving, while surviving great whites are moving elsewhere. Moreover, the fishery may also be illegally killing young great whites.

Though South Africa's Department of Agriculture, Forestry and Fisheries (DAFF), which regulates the nation's fisheries, gathers data on fish catches and populations, the department has yet to release specific data about current stock assessments of smaller sharks. And as of this writing, no scientific papers have been written that prove that these fisheries are wiping out sharks. To Fallows's great frustration, the South African government claims that the burden of proof is on him and other conservationists like Rochat, but he believes that the DAFF needs to prove that longline fishing *isn't* responsible for destroying shark populations. Regardless, by championing longline fishing jobs at the expense of the shark-tourism industry, which contributes tens of millions of dollars to the economy, the South African government is most likely sacrificing long-term economic stability for short-term benefits, because living sharks generate jobs and revenue. The shark cage-diving industry alone is worth R1 billion (US$70 million).[4] And yet despite the importance of sharks to the marine food chain, shark conservation remains a low priority for the South African government, which remains the only government in the world that promotes longline fisheries as an economic and employment stimulus program.

"Everything is connected," Rochat reminded me. "Since sharks play such an important role in keeping the marine ecosystems in balance, if the oceans are depleted of them, it will have severe repercussions on the millions of people who depend on the oceans for food. We need our sharks for these reasons, and for the simple joy of knowing that some of the most highly evolved animals in the world are still out there."

To protect sharks, Rochat told me, requires a steadfast support of conservation efforts by organizations and scientists, such as marine protected areas and shark sanctuaries, and a strict regulation of fishing practices, including catch limits, an outright banning of finning, and—perhaps most important, enforcement, even if it means prosecuting those who violate the law.

"Our greatest mistake," Rochat said, "has been to assume we are superior to all other life and to disconnect ourselves from nature, forgetting that we are, in fact, part of nature and that when we harm other life, we harm ourselves. Because everything is connected, we must save our sharks, so that we might save ourselves. We simply don't have the privilege of time anymore—nor do they."

Expanding on her "Rethink the Shark" campaign, Rochat launched a new campaign, this one called "Rethink the Predator," which publicized the deleterious effects of people on South Africa's dwindling shark population. Rochat and her team created a decal and posted it in aquariums around the world. Its headline warns: PREDATORS BEYOND THIS POINT. Below the warning is an information card that reads "Homo sapiens, a fierce predator found in both warm and cold waters, preys on sharks, skinning them alive, and leaving them to drown in the open seas. Offspring, if uneducated, may imitate behavior of adult species."

Seventy-five miles east of Cape Town, located in the beautiful resort town of Hermanus, is the headquarters of a group with a vastly different objective from that of Lesley Rochat. The nonprofit South African Shark Conservancy (SASC) seeks to understand the sustainable use of marine resources like sharks along much of the South African coastline. SASC is not a Greenpeace-type organization looking to put an end to illegal fisheries around the world; instead, the conservancy wants to figure out how fisheries can harvest sharks sustainably. As the oceans become depleted, SASC understands that people will turn to other species—like sharks, skates, or rays—to put food on the table. If this is as inevitable as it appears to be, SASC argues, it should be done in the most sustainable way possible to ensure that future generations can also benefit from the ocean.

The conservancy's location is awe-inspiring. Its 1,500-square-foot facility is embedded in a cliff overlooking a curved white-sand beach along the translucent green waters of the Atlantic. The wonderful smell of salt air filled my nostrils as I approached the entrance. Inside, I met with Tamzyn Zweig, a senior research scientist with the organization. Zweig started at the conservancy as a volunteer in 2010 and worked her way up to her current position as a manager for the conservancy's Sharks on the Line project, an ongoing collaboration with recreational catch-and-release shark fishermen. Recreational fishing is incredibly popular in South Africa, and according to the SASC website, it accounted for R2.5 billion (US$175 million) in revenue for the country in 2007, or about 80 percent more than the contribution of commercial fishing domestically.[5] However, this amount is less than half of the total value of South Africa's commercial fishery exports. Therefore, the government has an incentive to exploit fish resources like sharks. Zweig recognizes this tension and knows it's important to work with recreational fishermen to collect fishery data and to educate anglers about how to make sure caught sharks survive after they are released, because an indeterminate number of local fishermen catch and release sharks, skates, and rays without limit.

In 2010, the South African Shore Angling Association (SASAA), the governing body for competitive recreational rock and surf angling, expressed concerns about a perceived decline in the number of sharks caught, among other troubling trends. In response, SASC began tagging the sharks, skates, and rays caught and released at SASAA-sanctioned angling tournaments. Each tag recorded the length of the struggle, how long the fish spent out of the water, and, crucially, where the hook entered the fish. This information helps the conservancy determine what stresses individual sharks endure in the catch-and-release process. To date, SASC has tagged more than 2,000 sharks from twenty-seven different species.

"There is an organized angling sector where about 3,000 people catch and release sharks for sport," Zweig told me as we toured the SASC facility. "Each species will react differently to a very long fight time or a lot of air exposure on the beach. These people have to be educated that not all sharks are robust animals that you can drag around by their tails in the sand and that kind of thing. I don't think people harm them intentionally. As long as they are given the right information, I think it gives the shark a much better chance of survival once it's released back into the water."

I asked Zweig if she thought there was a way to teach humane catch-and-release techniques to fishermen. "There are a few things that catch-and-release shark fishermen can do to ensure the better health of the animal. Always choose heavier tackle to reduce fight time. If you're using really light tackle, it takes you a longer time to get the animal onto the beach. Once you've got the animal on the beach, make sure you work as quickly as possible to get the animal back into the water. If you want to take a photograph, or take a measurement, make sure that you're fishing with a friend who can help you get the animal back in the water as quickly as possible. And be prepared with a wet towel to put over the animal's eyes to reduce the stress in the animal until you can get it back into the water."

Zweig's advice brought me back to a time in my childhood when I was fishing with my dad. He hooked a fish, and the fight to reel it in was ferocious. I stayed on the bridge of the boat during the struggle, out of my father's—and harm's—way. A half hour passed, then an hour. My dad was in a chair and would reel in the line only to have the fish summon the energy to make a run again and take out all that line that my dad had just reeled in. This went on for quite a while. Two hours passed, and we were still not near the end. At last the captain threw the engines in reverse and took the boat to the exhausted fish, which was finally brought up to the stern of the boat. It was a marlin. I thought the fish should be let go and allowed to live after putting up such a brave fight. My dad wanted it, though, so he had the marlin stuffed and shipped home. That experience burned in me a respect for big game fish: his fight was courageous and a creature such as that should never be taken for vanity. Even if my dad had released the marlin, I learned from Zweig, it would probably have died anyway.

"When it comes to catch-and-release fishing," Zweig continued, "if you have caught a shark and it's swallowed the hook, it's very important that you don't pull that hook out, that you just cut it off as close as you can to the actual hook. The animal will reject the hook eventually. If you do pull it out, you're going to tear the gut, and that is almost certain death for the shark. Another thing you can do is choose where you're fishing. If you're fishing in an area where you know you're going to be using a lot of tackle and the shark is going to be cut off on the line, just bear in mind that that shark will be swimming around with all this line hanging out of its mouth. That line is likely to get entangled, and the shark will probably die from drowning."

While I believe it's possible—and necessary—for recreational anglers to practice a more humane catch-and-release style of fishing, I expressed my concerns to Zweig about commercial fisheries.

"South Africa is an industrialized fishing nation for sharks," she said. "We catch about 3,500 tons of sharks a year. The majority of that is exported to Australia and Spain, where it's sold under the guise of something else. So in Australia, you'll buy 'flake and chips' [fried fish and French fries] in a fast-food restaurant, but there is no mention of sharks. While people think they are eating fish, they're actually eating sharks." Few South Africans actually eat shark themselves, a great irony considering the near-extinction of the country's shark population.

"There's no international, even national, law that says that you have to name the animal to the species name," Zweig explained. "So you can call it a white fish, you can call it a blue fish. You need to ask the question at the place that you're buying the fish from.

What is this fish and how is it caught and where does it come from?"

Her view is that the South African government needs science to understand what is going on in fisheries. There is a dearth of data about shark fisheries. The government requires fisheries to report their catches, but when it comes to sharks, fishermen regularly misidentify the species or just report their catch as "shark." As a result, no one really knows what is happening to the populations of the various shark species. In addition, while many countries use observers, South Africa usually has no observers aboard its vessels. With limited government resources, only one or two fisheries are required by law to have observers aboard, and the shark fishery is not one of them.

"It's difficult to say whether there's overfishing," Zweig admitted, "but when it comes to the management of total allowable catches, the government believes that they have a good handle on the situation, but they really don't have good data to draw definitive conclusions. This issue is not just a South African problem; it is a global one. Overfishing from foreign nations in local waters has led to the collapse of many traditionally important commercial fish species, like your reef fish and that kind of thing. After this exploitation has happened, fisheries will turn to other species, like sharks. The government and the ruling party, especially, don't want to upset the citizens when they start taking away fishing quotas."

I asked Zweig if the South African government should do more to regulate commercial shark fishing. Without missing a beat, Zweig responded, "The government needs to make management of commercial shark fishing a priority. Sharks tend to be put on the sideline; they aren't considered as important as the other species. The South African government has the Department of Fisheries, which govern our fisheries, and we were appointed one shark scientist—only one."

"Doesn't that put tremendous pressure on sharks?" I asked incredulously.

Zweig told me she believes that the shark fishery is sustainable—as long as it is well managed. She admitted, though, that it's largely up to the individual fishermen to fish responsibly. "If the commercial fisherman goes out for the day and travels 100 kilometers to catch his fish and, unfortunately, he can't catch the fish that he's looking for, he is going to target sharks just to pay his bills." A few more years of that, and the entire South African shark population will be a thing of the past, putting into peril the country's maritime ecosystem.

"By nature, sharks are long-lived, slow-growing,

late-maturing animals who have very few young," Zweig said. "When you start adding pressure by removing these animals from the ecosystems, the effect can turn catastrophic very quickly."

Rochat and Zweig entertain divergent opinions about shark conservation. Zweig argues that by collecting data and educating recreational fishermen on how to catch fish, SASC can create an environment in which sharks can be sustainably harvested indefinitely. Rochat's view is that sharks are too valuable to the environment to be used as a resource. She believes that man has proved to be a poor steward of sharks and the environment. The two sides are at odds, and it is the role of the government to act as a fair judge. Just as the sharks bring balance to the ecosystem, perhaps the government should similarly bring balance between the competing demands of the commercial fishing industry and the conservationists. The latter group all too often gets ignored, but the revenues generated from maintaining natural resources like sharks can be significant. Ignoring the decline in sharks could lead to a disruption in revenue streams for the country. For that reason alone sharks should be protected. If the decline in sharks is allowed to continue, it is inevitable that it will ultimately hurt South Africa's shark-export business. And the damage won't stop there. The decline in sharks could lead to catastrophe in the marine ecosystem, affecting other commercial fishing and ecotourism.

Conservationists continue to work tirelessly to protect sharks by raising awareness of their plight. However, much needs to be done, such as stopping longline fishing and removing drum lines and shark nets installed along the beaches in the province of KwaZulu-Natal. The latter rack up casualties of great white sharks with as many as fifty deaths a year. There is always hope that the government might one day begin to listen to the conservationists like Rochat, but that day seems far off at the moment.

After our conversation, Zweig walked me over to a small circular shark tank, 4 feet deep with a 30-foot diameter. She wanted to show me Zeus, the conservancy's newborn pajama shark, a member of the catshark family that is indigenous to South Africa. Primarily nocturnal, the pajama shark, which has thick parallel stripes running up and down its body, lies motionless for most of the day, hidden in a cave. When threatened, it curls into a circle, covering its head with its tail. "The pajama shark is one of the 205-odd species that we have in our country," Zweig told me. "She's called a shy shark because of what she does."

Two months ago, the lab assistants noticed that one of the female pajama sharks had dropped what is known as an "egg case," little tendrils that hang out of the female. As the shark swims in and out of the kelp, the tendrils snag on the kelp, and the egg case nestles there for six to nine months, depending on the temperature of the water, until a baby shark is born.

I looked into the tank. Hiding near a rock was a miniature shark about an inch and a half long: baby Zeus, impossibly small and heartbreakingly vulnerable. "We don't handle Zeus," Zweig whispered, "because we don't want to stress her out. We want her to stay as healthy as possible. We'll just observe her and, when she starts looking stressed, we'll let her go just outside of the facility here. Basically, we're using her to show children and members of the community that not all sharks are big, scary, man-eating animals."

Someday, Zeus will be released into the wild and will join the fully grown pajama sharks that patrol the kelp-rich beds off Hermanus. She'll be part of a new generation of sharks taking their place in the South African waters, fighting for survival as they have done for eons.

The demand to create jobs is intense, and it is unclear whether the government will recognize the importance of protecting sharks to help the tourist economy and the marine ecosystem before the damage is irreversible. I would like to return to South Africa someday, but by that time I may find that dive experiences with sharks are exceedingly rare. By 2100, researchers say, climate change and overfishing will significantly threaten the existence of sharks.[6] At that point, swimming with sharks as Rochat has done might no longer be possible. People will have to know the great apex predators only through books. Rochat, Zweig, and the other shark warriors realize that if we lose the sharks, we stand to lose so much more.

Chapter 13
Shark Alley

The time had come for me to meet a great white in person.

Gansbaai is a fishing town and popular tourist destination in the Western Cape of South Africa. It is known for its population of great white sharks and as a whale-watching location. Tourists swarm to the town—which literally means "bay of geese" in Dutch/Afrikaans—for its lucrative cage-diving industry, which was started in 1995. Only Kruger National Park eclipses it as the go-to tourist destination in the country.

From Hermanus to Gansbaai is a short drive east. The road took me off the coast to avoid a nature reserve. Inland, the air is hot, the earth is dry, and vegetation is sparse. As I descended into the town, the ocean came back into view, and the luxuriant greens of the reserve stood out against the rocky gray beach. Waves were breaking far out to sea, rolling in toward the shoreline in rows. White ribbons of foam floated on pockets of royal-blue water. A cool breeze blew in gently from offshore. Along the water, in a bustling marina, nine cage-diving businesses were lined up below several life-size plaster statues of giant white sharks. Each operator was hoping that their oversize signs, like the blinking marquees on the Las Vegas strip, would draw in curious tourists eager to encounter the great whites in the infamous Shark Alley, the shallow channel between two islands: the larger Dyer Island, a nature preserve home to a declining colony of African penguins, and the smaller Geyser Rock, home to 60,000 Cape fur seals, which great whites nab as they traverse the channel.

My goal was to see a great white breaching out of the water in hot pursuit of a seal. I boarded a 35-foot powerboat at the dock and headed out to sea under the direction of Captain Nick, a no-nonsense man with a weather-worn face under his frayed baseball cap. As we motored out, he told me about growing up in the area and how his operation, which he started in 2008, brings in enough money to support his wife and three kids. A quarter mile offshore, we reached a nature preserve, a vista of deserted beaches with ancient 20- to

30-foot-high dunes and, beyond them, the green vegetation of rose hips and vines. The dunes pulsated with life: a dozen different animals burrowed in their white sand, leaving their footprints as a reminder that life can take up residence anywhere. Birds soared overhead and I could hear the seals yelping at each other. Under a blue sky, the water was a cloudy green. The spotter on deck called our attention to a dorsal fin at ten o'clock. We all turned at once, but the fin slowly descended from view like a periscope; most likely the shark was spooked or annoyed by the boat's engines.

As the boat slowed to a crawl, I spoke with Captain Nick about recent great white activity off Gansbaai. He told me that in mid-May, a few months before my visit, two notorious shark-eating orcas named Port and Starboard were spotted near Dyer Island, which caused the great whites to scatter to avoid their only predator. In 2017, a handful of 16-foot great white shark carcasses had washed up on the beaches in Gansbaai. The sharks' pectoral fins were still intact, but the middle of their bodies had been ripped open. The orcas had ganged up on the sharks and flipped them over, which put them into tonic immobility, essentially rendering them defenseless. The orcas then almost surgically removed each shark's oil-rich liver, which they devoured, and left the rest of the great white to rot on the beach.

While Captain Nick couldn't tell me whether the two orcas had recently killed any great whites—no shark carcasses were discovered in the channel—he admitted that the presence of great whites along the Southern Cape coastline was extremely fragile. This confirmed what Lesley Rochat and Tamzyn Zweig had told me. We stayed out for an hour and, during this time, not a single great white breached the surface. This made me worry even more that the dearth of shark sightings in Shark Alley signified something more troubling about great whites in South Africa.

Disappointed, I hopped in my car and left Gansbaai to continue my journey east along the coast. The drive was breathtaking. White sand beaches stretched for miles, and slow-breaking white waves rolled in on top of a light-blue sea. I had scheduled an interview with Enrico Gennari, PhD, a marine biologist and one of the world's leading experts on great white sharks. He lives and works in Mossel Bay, a thriving tourist town and headquarters of the Oceans Research, a marine and terrestrial research company. I had a hard time believing that this young man in front of me, with the dark hair and boyish good looks of an Italian movie star, was the company's director—until, that is, he started talking about great whites, which he's been studying for most of his life. "I was five years old the first time that

I watched a documentary on white sharks," he told me, "and I remember telling my mom that that was what I wanted to do. And, you know, for an Italian, it's usually quite strange because people in Italy become either a soccer player or an astronaut. So my mom said, 'Yeah, okay, whatever.' And instead of being a soccer player, I designed all of my career with that goal of being a shark scientist."

Gennari has tagged hundreds if not thousands of white sharks, and has collected more than a few fascinating stories. Once, he and his research team were conducting a study of the shark population off Seal Island, near Mossel Bay, on the Western Cape coast. Using sardines as bait, the team attracted four great whites, which circled the research vessel. From his port side, Gennari heard a splash. He turned to see a great white leaping out of the water. According to Gennari, the great white reached 10 feet in the air before it crashed down in his boat, right beside the vessel's fuel containers. Rather than letting the shark die onboard, Gennari and his team put a hosepipe into the shark's mouth to give it oxygen and worked to get it back in the water. They wrapped a rope around the shark's tail and tried to tow the shark into the water. But that attempt failed. They decided to head back to port, where the team used a crane to lift the 1,100-pound shark back into the water. Though the shark swam away, it couldn't navigate out of the harbor and soon beached. Gennari again tried to tow the shark out to sea. After an hour, the shark was able to swim away.

For all of his experiences with great whites, Gennari stressed—just like Greg Skomal—that we still have a lot to learn about the great white's biology and ecology. "We're still far from understanding this species, and that's what my research is moving toward."

Because Gennari has spent so much time with great whites, I thought that he could give me more insights into the species. Greg Skomal and Chris Fischer have uncovered a lifetime's worth of information about the great white's underwater behavior, but I wanted to learn more about what great whites are like in terms of personality. Gennari laughed and said, "White sharks are not clever, from a human point of view, but they're very inquisitive and curious. One thing I really figured out spending hours and hours with them is they are different. Not different from other species, but different among themselves. There are white sharks that are very curious. Others are more shy. And this is not true just about white sharks; it's well known in the animal kingdom. Everyone who's got a dog or a cat knows that two dogs are definitely not the same. So, the same thing applies to white sharks. Some travel much longer distances, while others really concentrate all their life closer to a specific environment. So there's no stereotype of the white shark."

While no one great white is like another, great whites do share many of the same characteristics as other shark species. Like any other species, for instance, great white pups are born alive and have to be ready to fend for themselves in an often harsh and unforgiving environment. Even though they are an apex predator, their survival at birth is a challenge, and their first test is getting food. "A young white shark up to three meters in length usually bases the majority of his diet on fish," Gennari explained. "And then, from three and a half meters, they change their diet to one more based on marine mammals—seals, elephant seals, et cetera. The thing is, to learn to hunt an elephant seal or a seal is quite difficult. And they don't have a mommy white shark to teach like a lion would do. So, there's no parental care in sharks. What they have to do is basically learning by making mistakes, and that is most likely the reason why it takes so long to be successful in hunting seals."

Gennari described the great white's triangular teeth, which can be as long as 2 to 3 inches. With about 300 teeth in their jaws, a shark's bite is formidable. Because a great white's teeth are serrated, they can tear through flesh. Other shark species, like the raggie, have blunt teeth like those of an alligator and can only grab and swallow their prey. All sharks lose their teeth regularly since they are not permanent and move forward gradually. A large shark can go through 30,000 teeth in its lifetime. But while small sharks lose their teeth every seven or eight days, great whites need several months to replace their teeth due to the time needed to develop these larger, serrated teeth.[1]

Gennari has been recently researching the behavioral ecology and physiology of great whites, specifically how they can raise their body temperature, body part by body part. "Every athlete knows that warmer muscle works better," he told me. "So white sharks, and the white shark family in particular, have a specific adaptation that allows the . . . red muscle to be warmer than the surrounding water temperature. In addition, the shark's stomach, the eyes, and the brain, can also be warmed up. Warmer means more effective, which is a huge advantage for a top predator like a white shark."

When we discussed the daily life of great whites, Gennari told me they are tremendous nomads. Like camels traveling in caravans across the desert, great whites traverse great distances through the sea. In fact, the travels of the great white make the peregrinations of other creatures pale in comparison. One female great white named Nicole traveled from South Africa to Australia in a straight line. Tagged near Gansbaai on November 7, 2003, she was detected about three months later when her tag—preprogrammed to release—floated back to the surface. Scientists retrieved it and, after reviewing the data, were convinced that the tag had malfunctioned or that a mistake had occurred. In less than a hundred days, Nicole had traveled approximately 6,800 miles, arriving in the Western Australian town of Exmouth, the longest known trip of any fish in the world. Though Nicole's trip was equal to the annual miles clocked by humpback whales during their migration, Nicole covered this distance in one-third of the time, averaging 70 miles per day. At the same time, Nicole traveled in a virtually straight line across the Indian Ocean along the 32nd parallel south, matching the migratory accuracy of other animal species, including whales, arctic terns, and seals. As Culum Brown and the group of scientists affiliated with the Charles Darwin Foundation established through their respective tagging programs, Port Jackson sharks and hammerheads can travel similar distances, returning to the same areas, often to the same reef, on return.

While scientists are only now beginning to understand how these animals make their journeys, Gennari and others point to the earth's magnetic field as a possible explanation for how these animals navigate. "Great whites maybe use the stars, maybe a magnetic field, maybe the current, maybe a mix of all those, but they are able to go in a straight line from South Africa to Australia and back, which is amazing," he said. "Me, without a compass—most likely even with a compass—I would never be able to reach Australia and back. I would get lost!"

A professor at UC Davis, Peter Klimley, told me that sharks use the earth's magnetic field to travel. Lava erupts on the seamounts just below the surface, and as the basalt oozes out, the magnetic particles line up along the north-south magnetic lines of the earth. With the lava creating valleys and ridges, the terrain features bicycle-like spokes that radiate out the magnetic information, which sharks use to gauge north-south direction during their sojourns of thousands of miles.

As we've seen, such migratory patterns leave sharks vulnerable to international fisheries. I wondered if, like the vulnerable hammerhead, the white shark's underwater travels have inadvertently put its population around the world in danger. "While it seems that the population is not in great shape," Gennari told me, "it's not in bad, bad shape. There are many more species that are really in trouble. White sharks? Let's say not in danger, but they're vulnerable." The International

Union for Conservation of Nature does, in fact, classify great whites as vulnerable, one step above the "endangered" category. But Gennari pointed to a few encouraging signs about the resilience of white sharks around the world. "One of the smallest white sharks ever recorded was caught in Turkey. There are still white sharks in the Mediterranean Sea, which is heartwarming for me, because I grew up in the Mediterranean. They are mostly unseen and are likely to be very low in numbers."

Still, vulnerable isn't safe or robust, and while people suggest different approaches to protecting South Africa's great whites, no one I spoke with denied that the great white population was in serious decline there. I thought back to Captain Nick, who relies on great whites to support his family. The next time I visit South Africa, I want to visit Nick again, and I hope his business will still be flourishing.

During the winter months, large shoals of sardines migrate into KwaZulu-Natal, a coastal province on the southeastern part of South Africa. To witness this sardine run, my plan was to travel to Durban, the province's largest city, an area known for its beaches, mountains, and savannahs populated by big game.

The sardine migration is one of the largest biomass

trips in the world, possibly even larger than the migration of the wildebeests.[2] The run is so vast that the shoal is visible from outer space. Sardines spawn in the cool waters of the Agulhas Bank and move northward along the east coast of South Africa. They come close to shore in the Durban area. The fish then turn northeast and eventually hit Mozambique, where they leave the coastline and, continuing farther east, eventually disappear into the vast blue of the Indian Ocean. The fish converge close to the shoreline because the cold currents along the coastline attract an abundance of plankton, which pull in the hungry sardines. And as we've seen before, where prey gather in great numbers, larger predators are sure to congregate. Dolphins, sharks, seabirds, and other species arrive in anticipation of the run, as if an internal alarm clock has gone off. These apex predators feast on the shoals of sardines. Divers can see the sardines and all the predators that take part in the feeding frenzy. Picture tens of thousands of birds soaring in the sky. The gannets patrolling the air look for dark spots in the water and hurl themselves downward. Tucking in their wings and turning themselves into missiles, they plunge deep into the clear blue waters for their prey. They zip through the water like arrows. And their beaks spear the sardines, which they snatch like prizes back to the surface. Bottlenose and common dolphins join in the frenzy, gorging on the tiny fish. The dolphins hunt as a group and surround the sardines, sending up bubbles to scare the little fish, who instinctively group tighter together into "bait balls." After a time, the sardines are so tightly packed that the dolphins can surge wildly into the ball and easily grab the sardines. Sharks patrol the area and lunge, snapping their jaws at the sardines. The numbers and variety of sharks at the sardine run are astounding: bronze whaler, Zambezies, hammerheads, coppers, and great whites by the hundreds. I had planned to hire a boat to witness the sardine run in person, but then I received some disturbing information.

When I was discussing my plans with Charles Maxwell, an Emmy Award–winning underwater cinematographer based in South Africa, he told me that I shouldn't bother, since there was nothing to see anymore. "It hasn't occurred for years," Maxwell said, to my dismay. "Two thousand and ten was the last very good year when I filmed whales feeding on sardines. Don't believe what you read on the internet or from boat operators about the sardine run. It's mostly commercial hype."

The principal cause of the end of the sardine run could be that the water temperature is rising in this area, just as water temperatures are rising in every other part of the world. As part of its fourth national climate assessment on global warming, the US government published the *Climate Science Special Report*,[3] which concluded that in addition to getting warmer, the oceans are rising and becoming more acidic because they have absorbed approximately 93 percent of the excess heat caused by greenhouse gas warming since the mid-twentieth century. "Ocean heat content has increased at all depths since the 1960s," the report noted, "and service waters have warmed by about 1.3°F since 1900 to 2016."[4]

The increase in water temperatures inevitably affects the sharks and the marine ecosystem. The annual sardine run is instructive. This event historically takes place when an offshore movement of the warm Agulhas Current is replaced by a cool, narrow band of inshore surface water, which provides the cold-water-loving sardine a nice cool corridor in which to migrate in large shoals. For the migration to take place, many believe the water temperature must stay below 69°F degrees. Since the water temperature is now higher, the sardine run has changed its normal course. In previous years, Maxwell tried filming the sardine run in Port St. Johns, where the water was colder. When the sardines failed to arrive there, he had to move again, this time farther south and west to East London, approximately 500 miles southwest of Durban, South Africa. Because East London is so far away from the sardines' traditional spawning grounds, the fish were less numerous. When the huge sardine shoals stopped migrating north, Maxwell told me, blacktip sharks migrated south to places like Aliwal Shoal in Natal, about 100 miles south of Durban. "When I started filming tiger sharks there, we hardly saw any blacktips. Now there are so many blacktips that they have had a negative impact on the tigers."

Blacktips and local dusky sharks would feast off the sardine run, but without the run, these sharks had a problem finding food. Whenever there is a catastrophic change to the environment, the rule of evolution is simple. A species has three choices: move, adapt, or die. Maxwell believes that it is no coincidence that blacktips are moving south to the Aliwal Shoal, where there are more fish, because it's the only real option they have. And their numbers are starting to overwhelm the water's resident tiger shark population, which is suddenly being outcompeted. So due to the web of connections, tiger sharks are suffering from the decline in sardines resulting from humans' effect on the climate.

Some argue that natural oceanographic cycles occur over time. Fossil evidence, for instance, suggests that the Pacific sardines have experienced, independent of fishing, regular boom-and-bust cycles. Over the last 1,700 years, sardine numbers have reached a low point every sixty years. However, others point to global warming, which is making it harder for the fish to reproduce. Either way, the changes to the ecosystem are dramatic. Moreover, the *Climate Science Special Report* maintains that "there is no convincing evidence for natural cycles in the observation record that could explain the observed changes in climate."[5] Therefore, it would be extremely difficult to say that the water temperature change affecting the sardine run is caused by anything other than human-generated global warming.

During my journey into the world of the hammerheads in Florida, I witnessed the impact of global warming on the species firsthand. I learned that hammerheads ate the schooling blacktip sharks that migrate during the winter from North Carolina into South Florida's waters. The blacktips, in turn, feed on the massive schools of mullet, anchovies, and sardines traveling along the Florida beaches. Stephen Kajiura of Florida Atlantic University has been studying this migration since 2011. After earning his master's degree in marine biology and later a doctoral degree at the University of Hawaii, he started conducting research at Florida Atlantic. Surveying the area overhead in a plane, he counted the blacktip shark migration, and he sometimes spotted a single group of 15,000. Kajiura simultaneously tracked the shark's migration, concluding that the sharks are sensitive to water temperature. The sharks prefer a water temperature in the range of 70° to 75°F. Due to global warming, however, the temperature of Florida's water increased by 3.6°F, which has changed the black-tip migration. The blacktips literally turn around when the water temperature isn't in their ideal temperature range. Kajiura calculates that the number of sharks migrating into South Florida has dropped by a third as a result.

Global warming is causing changes in animal behavior on land and at sea, and the ensuing trophic cascade is unpredictable. Logically, the sudden presence of a large number of sharks in Central Florida will probably upset the ecological balance there. Similarly, the sudden absence of an apex predator in South Florida could have a precipitous impact on the marine ecosystem there. "Blacktips sweep through the coastal waters and spring clean by weeding out the weak and sick fish," Kajiura explained. Without major reductions in greenhouse gas emissions, the *Climate Science Special Report* predicts, the increase in average sea temperature could reach 5°F or higher by 2100. If that happens, the damage to sharks and the marine ecosystem may be intolerable. One final observation that Kajiura made is that the change of the blacktip migration can also have an impact on shark bites: if more sharks stay in the cloudier waters of Central Florida, incidental bites may increase in that area.

Even though the sardine run was no longer a possibility, I still wanted to go to Durban to dive with a tiger shark, a longtime dream of mine. Unfortunately, civil unrest resulting in the murder of immigrants there caused me to cancel my trip to Durban. Instead, I returned to Cape Town, where there was one thing left to cross off my Shark Route itinerary: seeing a great white in person.

I headed back down to False Bay to schedule a dive with one of the town's many cage-diving operators. Before I completed my dive with a great white, however, I wanted to visit Muizenberg Beach, widely considered the birthplace of surfing in South Africa. The wide sandy beach curves round the eastern coast of False Bay and helps produce the long gentle breaks that surfers love.

Surfers still congregate at this beach in large numbers, including Joseph Krone, who survived a shark attack. A young man in his twenties with dark hair and a friendly, open face, Krone told me about his attack. We spoke over coffee one afternoon at an outdoor café in the harbor town of Mossel Bay. From our table, I could see the tranquil ocean glistening under a clear blue sky. Krone was wearing a simple striped T-shirt and spoke calmly about his experience. "I think it was about seven in the morning before the competition started. It was at the point in Jeffreys Bay, and there were quite a few people surfing."

Krone recalled that the sky was overcast that day, and the water was brown, which lent the atmosphere, in his words, a kind of eerie overcast. "But even then," he said, "it never crossed my mind about a shark or anything." After catching a few waves, he was paddling back out, taking a wide angle to avoid the break, which put him in deep water. About halfway, he decided to rest. "I put both my arms up on the board and was just sort of lying flat. That's when it happened. I just thought someone was tackling me or I did something wrong and so someone was trying to beat me up or something or playing a joke on me."

The shark first struck Krone with its nose, a simple investigative tap, but the strength of it sent Krone flying. "That was actually probably what saved me. If it had bitten straight away, it would have taken a piece straight out of my middle." Everything seemed to happen in slow motion, Krone told me, as he was tossed in the air. When he splashed back into the water, his surf-board crashed on top of him, shielding him from the great white, at least enough to absorb the shark's initial bite, which snatched a huge chunk of fiberglass from the board. The next thing Krone remembered was his leash getting pulled. "Luckily," he said, "it was a thin leash, and quite quickly it just snapped. If it was a thick leash, it could have dragged me along with the board."

Once free from the board, Krone finally saw the shark. "That was probably the most hectic part for me, just to see the intensity of the animal," he said, "the muscles just flexing. It was just whipping from side to side." Krone could see the board, probably broken, sticking out on either side of the shark's mouth.

Another surfer, Shannon Ainslie, paddled out to help Krone. Ainslie knew what he was doing. A few years earlier, he had survived his own attack by two great whites in East London. "He was the only guy that paddled to me. He offered me his board, which was quite a gesture because obviously for him to be in the water, it would be quite hectic. And he was very reassuring," Krone said. "If he wasn't there, I think I probably would have panicked. I probably would have really got terrified."

Krone and Ainslie made it to the rocks, where a

crowd had gathered. Two people, who had watched the entire attack from start to finish, told Krone how lucky he was, something that was becoming more and more clear to him as details of the attack began to sink in.

"I wasn't scared at the time because my body just went into automatic survival mode. I was just observing and not really feeling too much or thinking too much. For me, the most special part was just to watch the shark. To witness the power of it—I think that's where I gained the most respect for sharks and great whites."

The next day, Krone competed in the tournament. He wasn't scared, he told me, but he was aware like never before about a shark's power and presence. "For a surfer, the main danger that we all think about is the possibility of running into a shark. It's very much a battle of the mind, where you try not think about it. And that's where most of the fear is, just in your mind. I see sharks as beautiful animals. They're very graceful and mostly peaceful. I just think they're amazing animals. I've only got more respect for them because of the attack. I never thought of them in a worse way after the attack."

After talking to Krone, I couldn't wait to experience a great white shark in person. I hired a veteran of diving with great whites, a man named Rob Lawrence. How he became a shark-cage operator is quite a story. Born and raised in Cape Town, where his family has been fishing the waters since the late 1800s, Lawrence spent his childhood playing on the beach and catching fish. Volunteering with the white shark research project in nearby Gansbaai was the natural next step. As a junior research assistant, he helped tag smaller sharks around South Africa. Preternaturally curious, he later went looking for sharks in False Bay. He and a friend commandeered a 12-foot skiff to see what they could find near the shores of Cape Hangklip at the eastern end of the bay. To attract attention, Lawrence and his friend towed a life jacket behind the boat, unaware of or unconcerned with the number of great whites that were almost twice the size of their boat in the bay hunting for seals. By dragging the life jacket, Lawrence had unwittingly created a seal decoy, baiting some of the largest great white sharks in the world. After a while, they saw a small island. The aptly named Seal Island is home to an estimated 64,000 Cape fur seals—and nothing else. Rising no more than 20 feet above the high-tide mark, the 5-acre island is essentially a narrow granite outcrop with no soil or vegetation.

Great whites visit the island seasonally to prey on the seals. Traversing the island's waters for fish is dangerous for the seals, which is why the area is referred to as the Ring of Death. During the winter months, white sharks at Seal Island can feed easily on the newborn seal pups learning to swim in the treacherous waters. The "predator naive" pups don't know how to avoid sharks—or even that they should. Of course, the great whites also feed on the adults, but pups comprise 90 percent of their kill. Every creature born on this planet must learn how to survive in an often-inhospitable world. Young seals need time to learn survival skills, which makes them vulnerable to the experienced hunters that stalk them. Seals learn to improve their odds of survival by traveling in groups. The success rate of the sharks when hunting a group of seals is only 20 percent, but it rises to just over 50 percent when hunting seals that are swimming alone.

When hunting, a shark will patrol along the seafloor, peering up at the surface in search of the silhouette of a seal. Once it sees a target, the 2,000-pound shark will switch from a horizontal position to a vertical position and launch an attack from the depths like a cruise missile launched from a submarine, propelling itself toward the surface at 20 miles per hour. The safest place for the seal is near the shark's tail, because the shark cannot reach around fast enough to grab the seal. If the seal strays from the tail, however, the shark will be able to twist and snatch one of the seal's flippers.

The sharks' feeding ends when the seal mating season starts in summer. The great whites leave the island to hunt fish out at sea when the increase in water temperature brings in large migratory fish like yellowtail and skipjack tuna.

As Lawrence and his friend were towing the life jacket behind their boat, they finally caught the attention of a hungry great white, which immediately took off after the life preserver, skipping behind in the boat's wake. The shark soared out of the ocean, exposing the contrast of its ghostly white underbelly and the dark marine blue of its towering dorsal fin. The shark's reentry caused a white water explosion, which sounded to Lawrence like a cannon shot. They had witnessed what few people have ever seen: the breaching of a great white shark. Lawrence turned and looked at his buddy, whose eyes were bulging out of his sockets, just as Lawrence's were.

"I was totally blown away by what I saw," he told me later. "It was then that I knew I wanted to spend as much time as possible at Seal Island."

He started working two jobs to save enough money to buy his own boat. About that time, he met a young woman named Karen. Like Lawrence, she used to spend a lot of time outdoors as a child, camping in the African bush and hiking. Their courtship was a bit unusual; together they explored the bush in South Africa and Botswana.

Around the same time that he could afford his boat, he and Karen got married, and the newlyweds—along with Lawrence's buddy from the skiff—started a company called African Shark Eco-Charters. With Karen handling the administrative affairs, Lawrence started escorting tourists to see breaching great whites.

I drove to Lawrence's boat, which was docked in Simon's Town, about an hour's drive due south of Cape Town, at the northern end of False Bay. The boat was 35 feet long, with enough room to accommodate a small crew and—more important for the purpose of my trip—a shark cage in the stern. The cage had thick bars and one small aperture for the camera to peer out into the water. It was a beautiful, sunny day, but as we headed out to Seal Island, we encountered a fog bank, which made the journey surreal. Lawrence pulled back on the throttle to reduce the danger of collision. As I peered into the mist, I could make out a dolphin pod playing in the bow wave just off our boat. I also saw underwater disturbances, swirls and eddies of white water against the canvas of marine-blue water, though I couldn't tell if it was a seal, a dolphin, or a shark.

Just then the sun broke through the fog, revealing Seal Island's outcropping of dark granite rocks. The cacophony of 65,000 seals barking, calling out to one another, filled the air. The wind blew the pungent smell of seal droppings to our boat. Lawrence dropped anchor into the dark-green depths of the island's Ring of Death. He and his crew tossed chum over the side of the boat, the red swath of fish chunks scattered in contrast to the blue water. A crew member struck a two-by-four block of wood against the floor of the hull, hoping that the rhythmic thudding would attract the attention of a curious shark. Lawrence threaded a rope with decapitated tuna heads, still oozing with blood, and tossed it into the water as a final enticement. I knew full well that a great white shark was swimming, unseen, somewhere in my vicinity.

Then Lawrence lowered the cage into the water. It was now or never, my moment of truth. It was easy to think about swimming with a great white back in my apartment in New York City, but not so much with my knees bent over the boat's bow.

A litany of things that could go wrong started to cross my mind. I considered what would happen if the line connecting the shark cage to the boat broke, dropping the cage down into the depths. I would have to get out of the cage and swim to the surface, likely coming face-to-face with a great white. But then I realized that my imagination could come up with all kinds of fantasies. Now was the time to apply everything I learned along my journey and, despite my increasing fear, grab hold of my emotions and think rationally. I remembered the scientists and marine biologists I had interviewed. Their faces and voices were clear in my mind like a video montage: George Burgess at the International Shark Attack File telling me that sharks are not man-eaters and have no interest in hunting humans; Enrico Gennari reminding me that the great whites migrate away from the tourist-filled beaches in South Africa in search of fish; Greg Skomal, back in Cape Cod, assuring me that even on the beaches where *Jaws* was filmed, great whites are only after seals.

It is a mistake to ignore fear; survival in nature is dependent on it, and in a broader sense, fear is what keeps all ecosystems chugging along. I came to understand that the fear created by the wolves in Yellowstone and the tiger sharks in the seagrass beds of Australia were crucial to maintaining the ecosystem. Mike Heithaus told me that it was not the tiger sharks' killing huge numbers of sea turtles and dugongs that saved the ecosystem but the change they effected in the behavior of their prey. Fear allowed the sea turtles and dugongs to stay alive without decimating the seagrass and, in the end, preserved the entire marine ecosystem. Controlling fear is a sign of intelligence, a necessary step in our evolution. I zipped up my wetsuit and entered the cage.

A few minutes later, fully immersed in 50°F water, I peered through the too-thin bars of the cage into the green miasma. I could hear my own breathing and the dull sound of my heart beating under the wetsuit's skimpy layer of foamed neoprene. Finally, a great white emerged out of the void, grabbing at the decapitated tuna heads. Above me, on board, Lawrence and his crew pulled in the rope, jerking the line out of reach. The shark hurtled out of the sea with its mouth agape. It came crashing back down in the water near my temporary home, its weight causing a plume of white water to erupt, before it disappeared again into the green depths.

Another white shark, about 10 feet long, hovered into view. In no hurry, it meandered through the water with a gentle swish of its tail. As it circled the boat, the shark looked at me. I locked onto its eyes, which were coal black and seemed indifferent. I spotted no traces of aggression whatsoever, but I was moved by the stealth with which the shark surveyed the area around it, seemingly taking it all in at once. Because it didn't spot anything of interest—not me, a floating bipod in a submerged cage, nor the floating buffet of tuna heads—the shark swam away, entirely unimpressed.

Eventually, the tuna heads Lawrence and his crew continued to add to the line attracted a 20-foot great white. The massive shark circled in front of me and, with each turn, got nearer and nearer to my cage. Its presence was commanding, overwhelming, and inspiring, all at the same time, like witnessing a battleship go sliding by. I marveled at the shark's towering signature dorsal and pectoral fins, the battleship's forward and aft turrets. It was magnificent.

As the shark's pointed nose carved through the water, its black eyes exhibited a cautious curiosity. Suddenly visible, its arsenal of teeth, which have graced so many magazine and book covers, caught my attention. Instead of generating dread, its teeth inspired respect and awe.

My trepidation immediately gave way to excitement. I stood silently near the bars of the cage, in awe of the shark's strength and gracefulness. One of the ocean's fiercest predators was inches from me, a true emperor of the deep. An admiration and a sense of connectedness with sharks and all other ocean apex predators started to spring within me, a feeling of awe I can only describe as spiritual and pure. Absent old prejudices and unfounded fears, I recognized sea life as part of an integrated whole. As if wearing new lenses, I witnessed the true magnificence of the oceans and its inhabitants. My two-year journey was complete. I had experienced the full majesty of the shark. But even then, hundreds of feet below the surface, I realized there was still more to see, still more to experience, still more to learn, and still more I would never see. However briefly, I had traveled beyond the world of form into the mysterious and the eternal, stealing a glimpse of the species that has ruled the seas for 450 million years, a steadfast guardian of the world's oceans.

And just like that, the shark disappeared into the darkness.

Chapter 14
Save the Shark

To hold on to that feeling of excitement of having dived with a great white shark, I took one final trip, this time to the Bahamas, where I connected with an operation that offered cage-free dives. My hope was to finally swim in the open water with the elusive, or at least elusive-to-me, tiger shark. Prepared with a row of scuba tanks secured to its inside rail, the chartered boat headed out into the turquoise waters of the western Atlantic. After suiting up, I stood at the edge of the boat, peering once again into the blue depths below, finally ready to swim freely with the sharks.

Plunging feet-first into the water, I couldn't see anything because the bubbles from my oxygen tank obscured my vision. Eventually, the water cleared, and I stared down to the flat bottom, approximately 40 feet below my flippers. A shark was swimming along the seafloor. To get a better look, I headed down the anchor line with the excursion's dive master, my chain mail–adorned squire, who was carrying a cylinder filled with fish. The fish were for the sharks; the chain mail was for him, protection against an overzealous shark. I, on the other hand, was clad only in a wetsuit, completely exposed. Caribbean reef sharks swarmed around us, their snow-white underbellies suspended in the blue around us like clouds in the sky. Floating forms slowly took shape: bow-shaped mouths, broad noses, ocher eyes, pupils slit up and down like a cat's. Interested only in the dive master's fish, the sharks circled around him, lining up for their morning meal. The dive master skewered his spear with a fish and held it aloft. A reef shark snapped it up. More sharks quickly emerged out of the reef to join the party. A gray reef shark was coming right at me. The shark glided over my head, its tail swishing from side to side. I reached out and, ever so slightly, touched it. I know I shouldn't have, but I had to. The instinct was the same as wanting to pet a dog or cat or any beautiful animal, wild or domestic. Little did I know I was about to get all I could handle.

Grabbing another fish from his cylinder, the dive master fashioned it to his spear, then raised it about a foot in front of my face, not nearly enough room between me and the skewered bait for the shark to pass through without coming unnervingly close to my head. Out of the corner of my eye, I saw a reef shark charging. There was nothing that I could have done but remain motionless. The shark came in, considered me and the fish, and then chomped down on the bait, the right choice as far as I was concerned. As the shark raced by, the fish secure in its wide mouth, its pectoral fin grazed my cheek, and I felt the shark's smooth skin. (Shark skin is only rough against the grain; it's smooth going the other direction.) Sharks are covered with a slight film that protects them from bacteria and parasites, and this gelatinous residue marked my face like a calling card. I finally got my wish for a connection, literally.

I finished my dive and headed back to the boat, exhilarated. I tossed my fins and mask back on deck, and the dive master pulled me aboard. At no time was I in any danger. Instead, I was astounded by the sharks' remarkable beauty and the experience of being surrounded by so many of them. Although once again I failed to spot a single tiger shark, I no longer cared. The moment was thrilling, and sometimes it's best to leave a little mystery so we can appreciate all that remains to be discovered, especially in nature. Someday, perhaps, I will get to meet a tiger shark. I just hope it's not too late.

Sharks are facing a cataclysm over the next few decades if current trends continue. The international rescue mission currently underway—from Greg Skomal at the Massachusetts Shark Research Program to Lesley Rochat and her creative conservation efforts in South Africa—is about more than saving a specific species of shark. Explicitly stated or not, the ultimate goal is to protect every species, including humans. Combined, the efforts of scientists, conservationists, and environmental activists aim to safeguard the health of the ocean and the health of the planet, while simultaneously respecting and preserving the interconnectedness of every living species, of every living environment.

Over the planet's 4-billion–year history, this interconnection shaped a delicate balance of animals. In every ecosystem, apex predators maintain that balance crucial to the pyramid of life. If apex predators are harmed, trophic cascades reverberate throughout the system. The national parks do better when wolves are in abundance. The same is true for the oceans, where sharks maintain the health of the ecosystems. They patrol the high seas, keep the oceans clean, and watch over the coral reefs. Without sharks, the marine pyramid will crumble and wreak unforeseen consequences. Because sharks keep the oceans healthy, if we disrespect them, we disrespect the oceans. If we disrespect the oceans, we risk destroying the planet's primary source of life, laying waste to our own future as a species. Respect for life is the foundation for protecting the most important things in life. And that respect begins at the top of the marine food chain with sharks, the bellwether species for the health of the oceans and the working conditions of the fishermen who rely on the oceans for their livelihood and food.

I shared with Greg Skomal my concern that we're staring down a future in which entire regions of the world's oceans are bereft of sharks. "That's right," he said. "They've been fished out in many places. There could be remote atolls that don't have sharks anymore. The high-seas fleets that are targeting these sharks—I should add, illegally for fins—can move in and wipe the sharks out very quickly. And there's no one out there to enforce anything. And once the sharks are gone, they're gone forever."

To avoid this catastrophe, our mind-set needs to change.

Reasoning with people can help, but it has its limitations. The author Arthur Koestler argues, convincingly, that the voice of reason is up against an intractable foe. In *The Ghost in the Machine*, he writes:

> All efforts of persuasion by reasoned argument rely on the implicit assumption that homo sapiens, though occasionally blinded by emotion, is a basically rational animal, aware of the motives of his own actions and beliefs—an assumption which is untenable in the light of both historical and neurological evidence.[1]

To overcome this, Koestler says that we, all over the world, would need to experience a spontaneous, radical change in our consciousness. The twin undertakings of protecting the environment and saving sharks require such a change. Consciousness is not a set of opinions, information, or values but a total configuration in an individual that makes up his or her whole perception of reality or worldview. Included within the idea of consciousness is a person's education, politics, and values. But consciousness is much more than that, more, too, than their sum. It is that by which people create their own lives and thus the society in which they live. When civilization changes, the existing consciousness is likely to be in substantial accord with underlying material realities. However, when change is rapid, consciousness can lag increasingly behind reality, inadvertently or by willful ignorance. Over the past fifty years, we've experienced an accelerated rate of change that confounds our comprehension of it. As a result, we've seemingly lost our ability to recognize reality.

Up until recently, humans viewed the oceans—and their many resources—as infinite, indefinitely exploitable. Because the bounty of the ocean was inexhaustible, we could take fish at will. Sharks, once viewed as nuisance fish that got in the way of valuable commercial species, have been killed in unspeakable numbers. This worldwide slaughter was excusable because sharks have been considered dangerous man-eaters, underwater monsters to be feared and eliminated. Killing them for sport was as acceptable as was killing other apex predators like lions and wolves. Another commonly held belief was that jobs and money should always trump nature and conservation. Creating jobs and generating revenue justified damaging the environment irreparably.

And where has this historical consciousness led us? Society is slowly strangling the oceans by overfishing, dumping waste into the seas, discarding tons of plastic that span hundreds of miles in the ocean, and hurling mercury into the atmosphere that turns seawater into poison. Slowly, the oceans are losing the great apex predators: sharks, tuna, swordfish, and other fish that have lived in the oceans for millions of years, literally eons before humans arrived on the scene. The unfortunate new reality is that the oceans do not hold an infinite supply of fish to exploit. Because our actions are literally destroying the ocean environment, the old consciousness is therefore no longer appropriate.

The new consciousness wants to combine resources, capital, and time to maximize efficiency; there is no option that requires despoiling the environment so that people can pursue their livelihood. No inherent conflict exists between conservation and capitalism. Society can use reason and science to develop ways to produce goods and services without degrading the environment, which is now crucial for humankind. A rational approach is no more evident when managing renewable and nonrenewable resources. Gold, for example, is a nonrenewable resource. If you have a gold mine, you want to extract the gold as quickly as possible. The current approach to fishing sometimes treats commercial fishing as if it were a nonrenewable resource, which endangers the fishing industry and jobs. When fish are viewed as renewable resources, and managed as such, consumption of fish stocks can continue indefinitely.

The catastrophe of the cod bears witness to what

happens when renewable resources are not properly managed. In fact, the best way to make sure there are enough jobs for people in fisheries around the world is to make sure conservation succeeds, just as Tamzyn Zweig at the South African Shark Conservancy argues. This mind-set is what society needs, not antiquated beliefs supported only by shopworn arguments.

The new ecological consciousness has tremendous political implications. It recognizes that the welfare of the community and the world needs to be considered, and we must all work together to help create a better place to live.

Many Asian countries are still holding on to the old consciousness, which is why this transformation will take time. However, some countries have demonstrated that a change in perception *can* take place. As already discussed, Palau and the Maldives have created shark sanctuaries in their economic zones where commercial shark fishing is prohibited. Honduras also implemented a shark-fishing moratorium. After conservationists petitioned the Indonesian government to create a shark sanctuary near the island of Raja Ampat in western Papua, the Indonesian government passed legislation to create a 15,000-square-mile shark reserve. And in Fiji, conservationists are working to get better protection for sharks through wide-reaching media and education campaigns. These actions provide reasons to be optimistic that some nations recognize that the time has come to change course.

To be sure, changing the popular perceptions of sharks will be hard. Some nations will always view the species as an exploitable and expendable resource, and some individuals will always view killing apex predators as entertainment. While it is important to recognize that changing views is a great challenge, views nevertheless *do* change, more often than not by the sons and daughters of the future. There is still hope for sharks; the next generation offers it. The new generation coming of age wants to protect the environment, not exploit it. These emerging voices want laws to protect the planet for the future. They recognize that there is more to the world than the antiquated morals and ideas of previous generations. Millions of young people today are realizing that subtle harmonies and interrelationships exist among the species, and they have an underlying appreciation for the interconnectedness of life. In this new consciousness, humans are not here to exploit the environment. The relationship between people and nature is no longer antagonistic; it's synergistic.

In Florida, marine biologist and shark conservationist Jillian Morris is helping to educate and train the next generation of shark advocates through her Sharks4Kids program, which she launched in 2013. Through educational programs, field trips, and underwater adventures, she hopes to empower and inspire young people to appreciate sharks.

The sunny Morris, who resides in Florida with her husband, the underwater photographer Duncan Brake, designed an interactive, online curriculum for students and teachers to access, regardless of where they reside. She believes that children should use their voices. "We want to give them tools—both academically and out in the field—to encourage them, and for them to appreciate how amazing these animals are. No matter how young or old someone is, they can make a difference every single day." To date, the Sharks4Kids program has connected with almost 50,000 students in forty-seven different states and thirty-seven different countries, according to Morris.

In addition to the ocean- and shark-heavy curriculum, another key component of Sharks4Kids is collaboration. "We're big believers in the importance of science," Morris explained, "and we want kids to be exposed to all different elements, whether it's laboratory work or out in the field."

In Bimini, Sharks4Kids students regularly venture out into the mangroves, accompanied by certified program administrators, to interact with juvenile sharks up close. As part of the trip, students also visit the shark lab and facilities to learn more about the work that goes on there. In the Florida Keys, Sharks4Kids administers tagging programs with Seacamp, a program founded by Jeffrey Carrier, one of the world's foremost experts on sharks. As its name implies, Seacamp invites children to spend several weeks at various times of the year learning about the oceans and sharks in the Keys. Evenly split between girls and boys, groups of students work with Carrier to gain a general introduction to sharks and the threats that they're facing. Then each student gets a chance to perform a workup with a nurse shark. Over two days, they go out and catch and tag sharks as part of an ongoing survey. "Education is crucial in saving not only sharks, but the oceans in general," Morris said. "As adults, we get set in our ways, but younger generations haven't been inundated with the media. They haven't all seen *Jaws.*"

I asked Carrier if girls were unnerved by working with sharks. As soon as I asked the question, I knew the answer. "I don't think that the young women are at all scared off by sharks," said Carrier. "I think they bring the same attitudes as a young man would—that is, one of questioning. They haven't had any experience with the animals before, so they might be a little nervous to begin with. But the people who are participating with us in this program are successful women scientists. We have two classroom teachers from Monroe County schools who have a background in marine science and who have worked in marine science. Not just studied it, but they are marine scientists. We have our boat captains, our women boat captains. Other people who work with us in the program are successful women scientists, and I think that having role models that they can pattern their interests after is important.

"So," Carrier continued, "I think that there's just a little trepidation to begin with and maybe the fear of the unknown, but that goes away the first time they touch a shark. It's all of a sudden something very interesting to them when they learn they can handle the animal." At the end of the day, when the students realize that "they haven't been harmed, and they haven't harmed the animal," Carrier said, "we've got an attitude adjustment that's occurred."

Carrier told me that through new deep-sea DNA sampling technologies he and his teams are continuing to identify new shark species. "Common names get in the way sometimes," he said. "The bull, the cub, the ground, the Lake Nicaragua shark, and the Zambezi River shark all turn out to be the same species. And fortunately, modern genetics is allowing us to do the forensic work that can tell us a little bit more about that."

Despite the species' increased vulnerability, the total number of shark species now exceeds five hundred, a number that will continue to grow as sampling technologies improve. New species are usually found in deep water and possess extraordinary adaptation skills. For instance, scientists discovered the pocket shark in 1979, less than forty years ago. A team found a female specimen—the first of only two specimens taken so far—off the coast of northern Chile, 1,000 feet below the surface. The pocket shark is identified by two pockets next to its front fins, the purpose of which remains unknown. The pockets themselves, measuring about 4 percent of the shark's body length, are large, considering the shark's total body length of only 5 inches. Some researchers hypothesize that these pockets may secrete a kind of glowing fluid or pheromones. A more recent discovery in 2018 was a deep-sea shark named "Genie's dogfish" after the shark research pioneer Eugenie Clark. The shark lives in the Gulf of Mexico and the western Atlantic Ocean. Its blue-green eyes are large enough to see in great depths, but the shark itself only grows to 28 inches in length. As ocean exploration continues, scientists will likely discover new shark species, but that doesn't necessarily increase the species' chances of survival.

I asked Carrier about the outlook for sharks, and he

summed it up well: "One hundred million sharks are taken a year, and it's impossible to minimize the damage from that number of animals taken. There are still countries in the world that illegally fish for sharks and condone finning of sharks. The problem, of course, is that we're dealing with generally migratory species. They don't respect country boundaries, and the laws from country to country vary. And until we get together and have an international organization that can establish laws that are adhered to by fishing nations, I worry for our stocks down the road. I think that when you're dealing with any migratory species, whether it's a bony fish, shark, or bird, we must get an international agreement on management. . . . If such an international agreement is absent, I think these species don't stand a chance."

Carrier believes Americans should take the lead on the issue. "Mankind should go one step beyond that," he added. "Americans are the fortunate few in the world that can have an impact on other countries and their cultures," he said. "Every day, . . . what Americans buy, and what Americans talk about, has an impact around the world. And Americans should use that power to make sure that products they buy don't use slave labor, and that the environment is not destroyed to fill our plate. While Americans can look to the sky and dream great things, we still have to remember that the opportunity for change lies at our feet. We just have to take the first step."

The first step in preserving the oceans is developing a new consciousness. Reason and science can help ensure that proper regulations are in place to protect sea life. Similarly, educational programs like Morris's Sharks4Kids can also help protect the marine ecosystem and the lives of those men and women who work to provide our seafood in the long run. But the emergence of a new consciousness is essential if we are to have any chance of making significant changes and effecting real progress. When the right consciousness is in place, along with reason and education, the appropriate legislation—such as banning shark tournaments and controlling longline fishing—could automatically happen.

One need only look at killer whales for an example of how the culture can change its perception of an apex predator. At one time, not too long ago, people feared them, calling them the most savage of all animals in the sea. When they weren't reportedly swallowing porpoises and seals by the dozen, killer whales were rumored to knock men off icebergs to eat them. But then reality started to seep in. In 1965, a 22-foot-long male was captured near the town of Namu, British

Columbia. Today, almost everyone knows the story of this killer whale, which was later named Shamu. But back then, because Shamu was only the second killer whale captured, the species was still the stuff of nightmares: sea-dwelling homicidal maniacs, not at all the intelligent, sensitive creatures we think of today. In 1968, SeaWorld in San Diego opened its doors, putting Shamu on exhibition. As part of the show, the trainer brushed Shamu's teeth with a giant toothbrush, and then inserted his head inside Shamu's mouth. Everything people knew about killer whales was changing right there before their eyes. It was as if someone were trying to kiss a king cobra. Even before smartphones and texting, the new view of killer whales went viral. SeaWorld made stuffed Shamus that little children took home to cuddle with in their beds. More recently, the documentary *Blackfish* showed how killer whales like Shamu suffer in captivity in SeaWorld. If such a change in perception can happen for killer whales, it can happen for sharks.

Beyond the story of Shamu, the larger history of whales is instructive. Humans hunted whales almost to extinction. By the late 1930s, more than 50,000 whales were killed annually, and the world's whale population was closer to extinction than sharks are today. To save the whales, the world community decided to intervene.

The International Whaling Commission (IWC) banned whale hunting in 1986. Thirty years later, no whale species was brought to extinction, and many are now in the process of recovering. The same worldwide action can save the sharks. A similar "fins attached" law for sharks can help save sharks around the world just as the IWC's whaling ban did for whales. Following such an international decree, the sharks, the seas, and fishing industries will show immediate improvement.

As Jeffrey Carrier suggested, American consumers have tremendous influence. Seafood is relatively inexpensive across the country; cans of tuna regularly sell for less than the cost of the Sunday *Times*. But the low price of tuna carries a cost in terms of the long-term health of the ocean that does not show up on the final grocery bill. Taking into account fair labor wages for fishermen and canning employees—and the damage inflicted to the ocean through overfishing and the ecological impact of the slow annihilation of sharks—a single can of tuna isn't the bargain it's often made out to be. When consumers use their purchasing decisions as a form of protest and demand change, it sends a powerful message to producers, who live and die with the demands of the consumer. Consumers in general want to purchase sustainable food captured and harvested by fair-labor hands. Until recently, though, they have been unable to access accurate, up-to-date information about where their seafood comes from, how it's caught, and by whom. This is slowly starting to change, however, as exciting technologies are capable of bringing this information to average consumers.

John Amos is a pioneer in this kind of radical transparency, the technological next step in Greenpeace's subversive act of "bearing witness." Satellites in geosynchronous orbit over the earth can follow events in real time and allow people to see and understand the environmental consequences of human activity on the planet, including the oceans. In 2001, Amos left the corporate world to start a nonprofit called SkyTruth, which uses satellites to capture images of ocean and landscape degradation caused by mining, deforestation, and other human activities. In April 2010, when the *Deepwater Horizon* oil rig exploded, SkyTruth was the first to challenge BP's inaccurate reporting about the rate of oil spilling into the Gulf of Mexico. Amos and his team used satellite images to estimate the actual amount of oil gushing from the damaged well. Later, Amos realized that he could create a big data technology platform leveraging satellites to provide vast amounts of information that could be useful in regulating and monitoring fishing. Such a platform could create a truly global view of commercial fishing and might assist in improving the way fisheries operate.

Because he realized he couldn't undertake this project alone, he approached Google and the conservation advocacy group Oceana about partnering with SkyTruth. Google was a perfect partner because the company had already developed the Google Earth platform, which gives anyone with a computer the ability to view ships anywhere in the world's oceans. Anyone can go online and see whether any illegal or unregulated fishing is occurring. And because Oceana builds campaigns to implement science-based policies in countries that control one-third of the world's wild fish catch, they were a natural third partner. As one of the largest international nonprofit organizations in the world dedicated solely to ocean conservation, Oceana's missions include stopping illegal, unreported, and unregulated fishing, preventing habitat destruction, and protecting threatened and vulnerable species like turtles and sharks. Working with Google and Oceana, Amos and his SkyTruth team launched a program in 2014 to monitor fishing operations, called Global Fishing Watch (GFW), which uses satellites to track and detect commercial fishing anywhere around the world in real time. The goal of the program is to save the oceans from large commercial fishing operations that continue to overfish.

After learning about GFW, I drove to SkyTruth's offices in Shepherdstown, West Virginia, about 70 miles from Washington, DC, to talk to Amos. As he described how SkyTruth can leverage the existing satellite infrastructure to track fishing vessels, I began to appreciate the beautiful simplicity of the program's mission. The way it works is straightforward. Every ship, regardless of its country of origin, uses the automatic identification system (AIS), which broadcasts a radio signal from the vessel to satellites around the world. That signal registers the ship's identity and location. Since each vessel's AIS signal is unique, it is possible to follow a particular vessel around the world.

In his office, Amos showed me a digital map with thousands of tiny points of light. Each one represented a fishing vessel. Most of the lights were bunched along the equator, where the warm water helps foster greater fish stocks. Based on the number of lights on the board, I was amazed there are any fish left to harvest after humanity's mechanized assault on the oceans. Amos explained that a computer system using AIS can create a complete and continuous track of every vessel's movement, including its average speed. Fish caught by a specific ship can be traced from the exact point of capture all the way to port. Through SkyTruth's AIS-tracking program, distributors, processors, restaurateurs, buyers, and certification agencies around the world can now demonstrate seafood's provenance and how it's handled throughout the fishery supply chain.

By examining the AIS signal over time, observers can estimate what kind of fishing methods a boat is using. "When we look at the automatic identification system broadcasts that a ship sends out, we can reconstruct how it's moved out in the ocean," Amos said. "We can actually see when they've put nets and lines and hooks in the water to catch fish. And as a consumer who likes to know where his food comes from, this is very interesting to me because now it gives us the possibility that we could see where our fish comes from—not only what company it comes from or what boat it came off of—but where and when in the ocean that fish was caught."

If consumers want to know, they should be able to simply look up the information online. Amos believes that most people, if presented with the facts, would prefer to eat seafood caught in a way that helps preserve the environment and provide decent livelihoods for those who work in the industry.

SkyTruth's benefits go beyond identifying where

seafood comes from. These satellite capabilities allow governments, especially developing nations with limited resources, to monitor the extent and nature of fishing in local waters. These countries now have a tool to stop illegal fishing. Moreover, countries can now block imports of fish from vessels flagged in countries that fail to enforce fishery laws. By helping manage fisheries in a more sustainable way, satellite surveillance like the kind SkyTruth offers can help bring back fisheries under stress and thereby protect the livelihoods of the hundreds of millions of people around the world who depend on the ocean for food and income. For example, a fisherman in Belize, where trawl fishing is illegal, can use GFW to alert authorities when foreign trawling vessels enter the Belizean exclusive economic zone to fish under the cloak of darkness. And because the vessel-tracking data is open and freely available to the public—and the Belizean media—the government of Belize cannot easily ignore it. Such information could force decisive Belizean oversight. In countries like the Philippines, local authorities can use GFW to protect the fishing rights exclusively provided to municipal fishermen. If a large industrial vessel enters an area where fishing is strictly restricted by law to smaller artisanal boats, GFW can identify the offending vessel.

This technology can also help eliminate slavery at

sea. "Some of these crews don't see shore for a very long time," said David Manthos, the director of communications at SkyTruth. "By tracking the location of these vessels and understanding who they meet up with out on the open ocean, the system can shed light on this aspect of human trafficking and enslavement on the high seas."

In a recent study, Global Fishing Watch found at least two instances of fishing vessels staying at sea for more than five hundred days in 2015 and 2016. One Chinese vessel left the Port of Singapore, cut swiftly across the Indian Ocean, rounded the southern tip of Africa, sailed north to the equatorial Atlantic, and then circled and crisscrossed a stretch of ocean between West Africa and Brazil. While on the high seas, it rendezvoused with other vessels on three separate occasions, a possible sign that the Chinese vessel was transporting slaves. More than five hundred days later, it returned to port in Cape Town. This journey raised important questions. What are the conditions for a crew held at sea for five hundred days? Did these men volunteer for the journey, or were they being held against their will? Using GFW technology, governments can follow up on ships and crew to see whether violations are in fact occurring.

"With transponders on vessels that track where it's

going, you can see where it's been, even fishing in illegal areas," said Abby McGill, who works at the International Labor Rights Forum, a nonprofit organization
protecting workers' rights. "If it's been out to sea for
years at a time, without having gone back to port, then
that is a yellow flag they are using slaves."

This satellite technology works both ways. It also allows workers to stay in touch with the outside world.
"Those transponders now can also give workers basic
access to Wi-Fi. Almost everybody has a smartphone
today, even in the countries that we're talking about,"
McGill said. "If workers can have just one of these
devices on at sea that provides them some level of con-
nectivity, they have some connection to the outside
world. They can at least start to raise alarms if something bad is happening on that vessel."

Many want to avoid buying seafood caught using
slave labor, and so consumers should have the ability
to know the relevant information about the seafood on
their plate. The best place to start is by asking questions at the fish counter or restaurant. Is this fish wild-
caught or farm-raised? What country is this seafood
from? If it is wild, how was it caught? While some restaurants or grocery stores may not have the answers,
they will go to their seafood suppliers and get the
information if enough people ask. The reality is that

the technology exists today to create a system to provide that information to consumers, thanks to a relatively new government program.

In 2018, the US government implemented its Seafood Import Monitoring Program (SIMP), which requires importers of certain seafood products to maintain records to prevent illegal, unreported, and unregulated (IUU) or misrepresented seafood from entering the United States. The government implemented SIMP to protect food security and to ensure sustainability of the world's ocean resources. Any importer selling into the United States under SIMP is known as "the importer of record" and must report key data about their fish, such as the type of fishing gear used and the evidence of authorization to fish, in addition to other information. The data is loaded into the International Trade Data System (ITDS), which is the government's single data portal for all import/export reporting. The rule requires certain, though not all, priority species of seafood to be traced from the point of entry into the United States back to the point of harvest or production to verify whether they were lawfully harvested or produced. Those key fish on the list are sharks, tuna, swordfish, and shrimp, which are particularly vulnerable to IUU fishing and/or seafood fraud. The SIMP is not a labeling program, and the information is confidential. However, US sellers of seafood do have the capability to trace their seafood through this program. While the program is not yet consumer-facing, it has the potential to change the way we consume seafood. Individuals do have the ability to play a role in protecting the oceans. Engaged and informed consumers can make a huge difference with their pocketbooks. (NGOs like Greenpeace and the Environmental Defense Fund have published consumer guides to help with seafood selection.) Some may be willing to pay more for seafood that is sustainably caught, while others might not feel the same way. The market can decide.

This one last story also shows hope for the future. A Japanese-flagged longline vessel, *Kyoshin Maru*, fished the southern Pacific Ocean in 2018. The officers were Japanese, and the fishermen were Indonesians. Under orders of the captain and his officers, the crew caught sharks, slashed off their fins, and watched the sharks, still alive, sink into the black depths. The fins were stuffed in various holds. However, at the end of the voyage, they had a problem. How do you get shark fins from Hawaii to the Indonesian black market to sell them?

The three officers hatched a plan. The Indonesian fishermen legally disembarked at Pier 36 in Honolulu so they could fly back to Indonesia with the shark fins in their suitcases. Meanwhile, the three officers stayed on board and sailed back to Japan to avoid getting caught. Once the fins were in Indonesia, their payday would be $60,000 on the black market.

In their greed, they stuffed 962 shark fins into suitcases. In total, the cargo weighed nearly 200 pounds. When the Transportation Security Administration (TSA) officials lifted up the luggage, they noticed the bags were particularly heavy. Alarmed, TSA officials x-rayed the suitcases, which revealed triangular-shaped objects inside the suitcases. When they unzipped the luggage, dried fins of oceanic whitetip, bigeye thresher, and silky sharks spilled out on the floor. (In the past fifteen years, all three shark species have seen their populations drop between 80 and 90 percent.) All ten crew members were arrested.

The three officers and the owners of the fishing vessel were also charged because Hawaii passed a law making it unlawful for any person to possess, trade, or distribute shark fins. Hawaii is the only state in the union to have such a law on its books. The Hawaiians have a consciousness rooted in their history and cultural connection to the Pacific. The corporate owners of the fishing vessel face a $5.5 million fine, and the three individual defendants face a fine of $2.7 million. Aiding and abetting the smuggling of goods in the

United States carries a maximum term of twenty years' imprisonment.

The judge released the ten crew members after five days in jail and they went back to Indonesia.[2] They were just the mules acting under orders. Because the three officers never entered the United States, they were not arrested. They remain at large, but warrants for their arrest remain out. This story is a microcosm of the current status of fishing; to benefit a few men and companies, men from poor nations are snared into slavery and ordered to commit atrocities that damage the marine ecosystem.

If Hawaii can pass legislation that protects sharks, why can't Congress? Important legislation is currently pending in the United States Congress. The Sustainable Shark Fisheries and Trade Act would require all countries importing products related to sharks, rays, and skates into the United States to obtain certification by the National Oceanic and Atmospheric Administration. This certification would require proof that conservation management and enforcement are comparable to those in the US. While shark-finning is illegal in US waters, shark fins continue to be bought and sold throughout the country. This bill will limit the supply of these products in the US to those that are certified and meet the stringent criteria contained in the bill. Rep. Daniel Webster (R-FL) introduced the bill as a way to ensure overseas fishermen are held to the same science-based regulations as fishermen in America. He said this bill "encourages other nations wishing to export shark products to the United States to adhere to the same high standards for conservation and management."[3]

More important, the Shark Fin Sales Elimination Act, introduced by Rep. Gregorio Kilili Camacho Sablan (I-MP) and Rep. Michael McCaul (R-TX), would make it illegal to buy or sell shark fins or any product containing shark fins in the United States. Every year, fins from 73 million sharks enter the global shark-fin trade, and the United States still participates in the shark-fin trade. Shark fins are imported into the US from countries that do not have similar shark-finning restrictions. Since 2010, the United States has imported fins from eleven countries, five of which do not have any type of ban on finning.[4]

Shark tournaments should also be banned in the United States. If the US is going to be the leader in ocean conservation and protecting sharks, the country must go beyond passing legislation and stopping its own form of "American finning," which is killing sharks as entertainment. The United States is sending the wrong message to the world in allowing such barbarous practices to continue.

The IUCN created the Convention on International Trade in Endangered Species of Wild Fauna and Flora (CITES), a multilateral treaty, of which the United States is a signatory, to protect endangered animals. CITES is one of the largest and oldest conservation agreements in existence. This treaty has listed twelve sharks that the IUCN views as either endangered or threatened and where hunting them bears a substantial risk to the species. Some of the sharks on the list—hammerheads, threshers, and porbeagles—are routinely killed in shark tournaments. At the same time, the tournaments contravene the scientifically based conservation measures of IUCN. It is already illegal to kill great white sharks in US territorial waters. Adding the same protections to makos, threshers, and other threatened species is consistent with current law.

In addition, these tournaments result in the loss of apex predators to the detriment of the marine ecosystem. Moreover, sharks are valuable assets that can benefit the local economy in ecotourist operations.

Where is the proof? South Africa has a burgeoning cage-diving industry that generates money and jobs from sharks. For example, Gansbaai, South Africa, attracts people from all over the world, and those tourist dollars pour into the economy from flights, hotels, and numerous other activities. Dozens of charter boat operators take scores of tourists on daily trips. There is only one shark cage-diving operation in Montauk. When I went on the Montauk cage dive, I saw makos and blue sharks. The trip was an engaging experience, but more important, it was good for the local economy. I spent money for a hotel, for meals, and for the tour itself. It was well worth it. With proper advertising, there could be several cage-diving operations in Montauk and other locations along the East Coast. It does not take a lot of imagination to realize the vast number of jobs that would be created to run these operations, maintain the boats, and provide ancillary services like meals, lodging, and other concessions.

With a new consciousness, we will once again feel a connection to the outdoors, the ocean, and its vast sea life. People who seek out those experiences with the sea will derive a great source of satisfaction. The salt in the sea is the same salt in our blood. The freedom of the sea lives in our breasts. Swimming in the ocean brings renewal and happiness. Nature is not something foreign but something that lives within us. We need this connection as a way to survive and fill our soul, otherwise we run the risk of the soul withering up like dried seagrass on the sand. Our spiritual survival depends on a recognition that we are part of the web of life and the ocean.

My two-year journey took me to some of the most beautiful places on earth, sharing the underwater environment with the sharks, who let me into their world, even as an uninvited observer. On land, I met with the world's leading scientists and marine biologists and today's most impassioned and creative conservationists and activists, and they taught me more than I thought possible about sharks and what it means to share the world with them. With all journeys, the traveler is changed. Perhaps I gained a few wrinkles from the sun, but I have a great appreciation for the interconnectedness of all life. As I listened to the breaking waves and heard what the ocean had to say, my connection with the ocean and marine wildlife grew. And the sharks were patient and taught me about life in the ocean. At the same time, I learned of the tremendous suffering by humans and animals that takes place on the high seas.

Through my journey, I have become an ocean conservationist.

I know there is a long road ahead to raise awareness about the massacre of sharks around the world, from shark tournaments in Montauk to the black markets of Beijing. But contests that require all your effort are really the only ones important enough to take on. The fight to save the sharks and the ocean will not be easy, but those who make the attempt will join in a fight that is crucial to humanity.

In fifty years, our children will stand on the shoreline and gaze out at the sea. What will they see? Will they witness a lifeless sea filled with broken reefs, amputated versions of the reefs' former selves, and dull, gray seafloors? Will the schools of hammerheads be only a distant memory? Will our children know the great apex predators only through dusty books on library shelves? Or will our children stand on a beach and see, unfolding before their eyes, the miracle of life? Will they hear the seabirds and the lapping of the waves, and feel the ocean breeze coming from a sea teeming with life? If we act today, we can ensure that our children will look out on the ocean and know that underneath the surface, a shark will be swimming, standing guard over its dominion, protecting the ocean, the greatest miracle on earth, just as it has done for more than 450 million years. And when our children close their eyes and dream of the countless wonders waiting to be discovered in the sea, they will know the full majesty of life and the mysteries it continues to offer.

Acknowledgments

O ver my journey I had the pleasure to meet with many remarkable scientists who have uncovered the inner workings of sharks and their world. Their research, based on years of study out on the ocean, often under challenging conditions, contributes to society's crucial knowledge of the environment to implement the right regulations to protect the oceans.

I started my journey with meeting Greg Skomal, who has worked for years on great white sharks in the Northeast. He was so generous with his time, and his insights were extraordinary. I also had great help from Jeffrey Carrier, for his insights and for opening up his Rolodex and introducing me to other marine biologists and scientists such as Mike Heithaus and his colleague Wes Pratt. Mike's study on sharks and the marine ecosystem in Australia is truly remarkable research. I enjoyed talking with him and with Wes, whose enthusiasm about sharks is infectious.

I have also had the assistance of so many others, like Tristan Guttridge, who has uncovered the social world of sharks, and Neil Hammerschlag, who has published a litany of important research reports. Pete Klimley has made likewise great discoveries, too numerous to mention here, over his illustrious career. Jelle Atema was another great scientist who was so generous with his time. He provided the sounding board for so many concepts that I wrote about. I wish I had more room so I could discuss more about each one of these scientists, their work, and how generous they were with me.

One of the key lessons along my journey is that everything is interconnected. A book like this can appear only because of the work of many others. When Paul Rachman, who has helped me so much along the way, introduced me to Peter McGuigan of Foundry Media, the book started to take shape with his great team of Richie Kerns and Mike Nardullo.

Working with HarperCollins was a dream come true. I want to thank my publisher, Judith Curr, who continues to champion the book, and Miles Doyle, a remarkable editor who put his heart and soul into this project. I also want to thank Suzanne Quist and her production team for turning the manuscript into a book.

To all these great men and women, I thank you and I am the first to say, this book would never have been written without you all.

Notes

Introduction: Man Bites Shark

1. John A. Musick and Susanna Musick, "Sharks" in *Review of the State of World Marina Fishery Resources*, Fisheries and Aquaculture Technical Paper 569 (Rome: Food and Agriculture Organization of the United Nations, 2011), 245–53, www.vliz.be/imisdocs/publications/ocrd/236437 .pdf.

2. Allison Pohle, "Why the Author of *Jaws* Wished He Never Wrote It," Boston.com, June 19, 2015, https://www.boston .com/culture/entertainment/2015/06/19/why-the-author -of-jaws-wished-he-never-wrote-it.

3. American Association for the Advancement of Science, "Longest-Lived Vertebrate Is Greenland Shark: Lifespan of 400 Years," Science-Daily, August 11, 2016, https://www .sciencedaily.com/releases/2016/08/160811143218.htm.

Chapter 1: Searching for Mary Lee

1. OCEARCH, "OCEARCH video: Tagging Mary Lee," video, 3:54, July 21, 2017, published by WTOP Radio, https://www.youtube.com/watch?v=WdHtlvHIcGE.

2. OCEARCH, "Naming Mary Lee," video, 2:06, February 19, 2013, published by OCEARCH, https://www.youtube.com/watch?v=wJ8pKZa4wZE&feature=youtu.be&t=17.

3. WAVY TV 10, "Interview with OCEARCH Founder Chris Fischer," video, 12:06, January 23, 2013, published by WAVY TV 10, https://www.youtube.com/watch?v=U6QH8Aa6VLA&feature=youtu.be&t=35.

4. Bioluminescent light is more visible at night without interference from sunlight. The best opportunity for humans to see bioluminescence in the ocean is in the evening when marine plankton sparkles in disturbed water, such as in the wake of a boat.

5. "Great White Sharks Are Taking Tropical Winter Holidays Too!," NIWA, March 31, 2010, https://www.niwa.co.nz/news/great-white-sharks-are-taking-tropical-winter-holidays-too.

6. "New Study Finds Extreme Longevity in White Sharks," Woods Hole Oceanographic Institution, January 8, 2014, http://www.whoi.edu/news-release/LongevityWhiteSharks.

7. Michael L. Domeier and Nicole Nasby-Lucas, "Two-Year Migration of Adult Female White Sharks (*Carcharodon carcharias*) Reveals Widely Separated Nursery Areas and

Conservation Concerns," *Animal Biotelemetry* 1, no. 2 (2013).

8. Mark Gomez, "San Jose Man Convicted of Killing Great White Shark," *Mercury News*, January 25, 2019, https://www.mercurynews.com/2019/01/24/san-jose-man-convicted-of-killing-great-white-shark.

9. "Structure and Function of the White Shark Brain," Biology of Sharks and Rays online course, n.d., http://www.elasmo-research.org/education/white_shark/structure_brain.htm.

10. "Structure and Function of the White Shark Brain," Biology of Sharks and Rays online course.

11. "Structure and Function of the White Shark Brain."

12. August G. Domel et al., "Shark Skin-Inspired Designs That Improve Aerodynamic Performance," *Journal of the Royal Society Interface* 15, no. 139 (February 2018), https://royalsocietypublishing.org/doi/full/10.1098/rsif.2017.0828.

13. Domel et al., "Shark Skin-Inspired Designs."

14. G. Del Raye et al., "Travelling Light: White Sharks (*Carcharodon carcharias*) Rely on Body Lipid Stores to Power Ocean-Basin Scale Migration," *Proceedings of the Royal Society B: Biological Sciences* 280 (July 17, 2013), doi:10.1098/rspb.2013.0836.

15. "Voyage to the White Shark Café," Schmidt Ocean Institute, https://schmidtocean.org/cruise/voyage-white-shark-cafe.

16. Salvador J. Jorgensen et al., "Eating or Meeting? Cluster Analysis Reveals Intricacies of White Shark (*Carcharodon*

carcharias) Migration and Offshore Behavior," *PLOS ONE* 7, no. 10 (October 29, 2012), http://whitesharkcafe.org /Publications/Jorgensen_2012.PDF.

17. David Freestone et al., *World Heritage in the High Seas: An Idea Whose Time Has Come*, World Heritage Reports 44 (Paris: United Nations Educational, Scientific, and Cultural Organization, 2016), http://whc.unesco.org/document /143493.

Chapter 2: Makos, the F-35 of Sharks

1. Guido Dehnhardt et al., "Hydrodynamic Trail-Following in Harbor Seals," *Science* 293, no. 5527 (July 6, 2001), 102–4, doi: 10.1126/science.1060514.

2. Jayne M. Gardiner, Jelle Atema, Robert E. Hueter, and Philip J. Motta, "Multisensory Integration and Behavioral Plasticity in Sharks from Different Ecological Niches," *PLOS ONE* 9 (April 2, 2014), e93036.

3. Tanya Brunner, "The Sixth (and Seventh) Sense," Shark Savers, n.d., http://www.sharksavers.org/en/education/bi ology/the-sixth-and-seventh-sense.

4. Author interview with Jelle Atema, January 14, 2019.

5. R. Douglas Fields, "The Shark's Electric Sense," *Scientific American* (August 2007): 78.

6. "Intelligence," Shortfin Mako Shark, n.d., https://ipfs.io /ipfs/QmXoypizjW3WknFiJnKLwHCnL72vedxjQkD DP1mXWo6uco/wiki/Shortfin_mako_shark.html.

7. Francis G. Carey, John M. Teal, and John W. Kanwisher, "The Visceral Temperatures of Mackerel Sharks (Lamnidae)," *Physiological and Biochemical Zoology* 54, no. 3 (July 1981): 334–44.

8. Jeanine M. Dooley et al., "Convergent Evolution in Mechanical Design of Lamnid Sharks and Tunas," *Nature* 429, no. 6987 (June 2004): 61–65.

9. "New Reports Highlight Landings, Value and Economic Impact of U.S. Fishing," NOAA Fisheries, December 13, 2018, https://www.fisheries.noaa.gov/feature-story/new-re ports-highlight-landings-value-and-economic-impact-us -fishing.

10. Jerome Kelley, "Fishing the Man-Eaters," *Yankee*, July 1974.

11. Frank Mundus, *Sport Fishing for Sharks* (New York: Macmillan, 1971).

12. Doug Criss and Samira Said, "Three Florida Men Charged in Shark-Dragging Video," CNN, December 13, 2017, https://www.cnn.com/2017/12/13/us/shark-dragged-video -arrests-trnd/index.html.

13. Craig Pittman, "Shark-Dragging Video Case Results in Three Arrests," *Tampa Bay Times*, December 15, 2017, https://www.tampabay.com/news/environment/wild life/Shark-dragging-video-case-results-in-three-arrests _163541187.

14. Jim Turner, "Gov. Rick Scott Asks for Rules Review After Shark Video," *Pensacola News Journal*, July 29, 2017,

https://www.pnj.com/story/news/politics/2017/07/29/scott
-asks-rules-review-shark-video/104108252.

15. Alan Lowther and Michael Liddel, eds., *Fisheries of the United States, 2013*, NOAA, September 15, 2014, https://www.st.nmfs.noaa.gov/Assets/commercial/fus/fus13/FUS 2013.pdf.

16. Austin Gallagher, Neil Hammerschlag, Andy J. Danylchuk, and Steven J. Cooke, "Shark Recreational Fisheries: Status, Challenges, and Research Needs," *Ambio* 46, no. 4 (May 2017): 385–98.

17. D. S. Shiffman et al., "Trophy Fishing for Species Threatened with Extinction," *Marine Affairs and Policy* 50, no. 1 (July 2014): 318–22, https://doi.org/10.1016/j.marpol.2014.07.001.

Chapter 3: The Mysterious Case of the Hammerhead

1. Douglas D. Lim et al., "Phylogeny of Hammerhead Sharks (Family Sphyrnidae) Inferred from Mitochondrial and Nuclear Genes," *Molecular Phylogenetics and Evolution* 55, no. 2 (May 2010): 572–79.

2. Lim et al., "Phylogeny of Hammerhead Sharks."

3. D. M. McComb, T. C. Tricas, and S. M. Kajiura, "Enhanced Visual Fields in Hammerhead Sharks," *Journal of Experimental Biology* 212 (2009): 4010–18, http://jeb.biologists.org/content/212/24/4010.

4. "Researchers Investigate Fishy Sense of Smell with Hammerhead Shark Model," ScienceDaily, May 6, 2010, https://www.sciencedaily.com/releases/2010/05/10050509 2521.htm.

5. Nicholas L. Payne et al., "Great Hammerhead Sharks Swim on Their Side to Reduce Transport Costs," *Nature Communications* 7 (July 26, 2016): 12289, doi: 10.1038 /ncomms12289.

6. In ancient Greece, dentists would extract the venom from stingray spines for use as an anesthetic.

7. Cesar Peñaherrera et al., "Hammerhead Sharks of Galápagos: Their Behavior and Migratory Patterns," in *Galápagos Report 2009–2010* (Ecuador: Charles Darwin Foundation, 2010), https://www.researchgate.net/publica tion/277813194_Hammerhead_sharks_of_Galapagos _their_behavior_and_migratory_patterns.

8. Patricia Beller, "Scalloped Hammerhead," Ocean Oasis Field Guide, n.d., https://www.sdnhm.org/oceanoasis/field guide/sphy-lew.html.

9. "Shark Fin Trading in Costa Rica," Wikipedia, May 14, 2018, https://en.wikipedia.org/wiki/Shark_fin_trading_in _Costa_Rica.

10. David Boddiger, "Shark Fin Mystery Deepens," Pretoma, accessed January 4, 2011; "Costa Rica Prepares to Export 10 Tons of Hammerhead Shark Fins," *The Costa Rican Times*, March 2, 2018, https://www.costaricantimes.com /costa-rica-prepares-to-export-10-tons-of-hammerhead -shark-fins/59434.

11. Janet Raloff, "New Estimates of the Shark-Fin Trade," *ScienceNews*, November 1, 2006, www.sciencenews.org/blog /food-thought/new-estimates-shark-fin-trade.

12. James Welsch, *Sharks Get Cancer, Mole Rats Don't: How Animals Could Hold the Key to Unlocking Cancer Immunity in Humans* (Amherst, NY: Prometheus Books, 2016).

13. Douglas Main, "Sharks Do Get Cancer," LiveScience, December 3, 2013, www.livescience.com/41655-great-white -shark-cancer.html.

14. Charles Lu et al., "Chemoradiotherapy with or Without AE-941 in Stage III Non-Small Cell Lung Cancer: A Randomized Phase II Trial," *Journal of the National Cancer Institute* 102, no. 12 (June 16, 2010): 859–65, https://doi .org/10.1093/jnci/djq179.

15. "Shark Cartilage Supplement Does Not Extend the Lives of Lung Cancer Patients," National Center for Complementary and Integrative Health, June 15, 2007, https:// nccih.nih.gov/research/results/spotlight/061507.htm; "Powdered Shark Cartilage for Advanced Breast and Colorectal Cancer," National Center for Complementary and Integrative Health, July 1, 2005, https://nccih.nih.gov /research/results/spotlight/sharkcartilage_rr.htm.

16. Douglas Main, "Sharks Do Get Cancer: Tumor Found in Great White," *Scientific American*, December 5, 2013, https://www.scientificamerican.com/article/sharks-do -get-cancer-tumor-found-great-white.

17. American Association for the Advancement of Science, "Sharks DO Get Cancer," *Science* 288, no. 5464 (April 14, 2000), doi: 10.1126/science.288.5464.259d.

18. "Cancer: Prevention and Early Detection," WebMD, accessed March 7, 2019, https://www.webmd.com/cancer/cancer-prevention-detection-18/default.htm.

19. Austin J. Gallagher, "Physiological Stress Response, Reflex Impairment, and Survival of Five Sympatric Shark Species Following Experimental Capture and Release," *Marine Ecology Progress Series* 496 (January 2014): 207–14.

20. "SNAPPED! Tiger Shark vs. Hammerhead Shark," Barcroft TV, July 27, 2016, https://www.youtube.com/watch?v=x0t5Rx30Qec.

21. S. M. Chavey, "Corpus Christi Man Makes Hammerhead Catch of 'Multiple Lifetimes' at Padres Island National Seashore," *San Antonio Express-News*, July 16, 2018, https://www.mysanantonio.com/lifestyle/travel-outdoors/article/Corpus-Christi-man-catches-14-foot-hammerhead-13079145.php.

22. "Hammerhead," IUCN Red List of Threatened Species, Version 2018-2, International Union for Conservation of Nature and Natural Resources, 2019, https://www.iucnredlist.org/search?query=hammerhead&searchType=species.

23. "Hammerhead Gaze," bioGraphic, December 13, 2017, California Academy of Sciences, https://www.biographic.com/posts/sto/hammerhead-gaze.

24. "Species Implicated in Attacks: 1580–Present," International Shark Attack Files, Florida Museum of Natural History, August 20, 2018, https://www.floridamuseum.ufl.edu/shark-attacks/factors/species-implicated.

25. "Angler Participation," American Sportfishing Association, accessed March 7, 2019, https://asafishing.org/facts-figures/angler-participation.

Chapter 4: Sharks as Social Animals

1. Bruce Friedrich, "Fish Are Smart (and of Course They Feel Pain!): An Interview with Dr. Culum Brown," *HuffPost*, August 31, 2014, https://www.huffingtonpost.com/bruce-friedrich/fish-are-smart-and-of-cou_b_5545914.html.

2. Culum Brown, "Fish Intelligence, Sentience, and Ethics," *Animal Cognition* 18, no. 1 (January 2015): 1–17, doi: 10.1007/s10071-014-0761-0.

3. Steven G. Wilson, "Basking Sharks (*Cetorhinus maximus*) Schooling in the Southern Gulf of Maine," *Fisheries Oceanography* (June 20, 2004), https://doi.org/10.1111/j.1365-2419.2004.00292.x.

4. Joseph Long et al., "Demography and Movement Patterns of Leopard Sharks (*Triakis semifasciata*) Aggregating Near the Head of a Submarine Canyon Along the Open Coast of Southern California, USA," *Environmental Biology of Fishes* 96, no. 7 (July 2013), doi: 10.1007/s10641-012-0083-5.

5. R. Glenn Northcutt, "Elasmobranch Central Nervous System Organization and Its Possible Evolutionary Sig-

nificance," *Integrative and Comparative Biology* 17, no. 2 (May 1, 1977): 411–29, https://doi.org/10.1093/icb/17.2.411.

Chapter 5: The Quest for the Tiger Shark

1. Erich K. Ritter, "Fact Sheet: Tiger Sharks," Shark Info, June 4, 2016, http://www.sharkinfo.ch/SI4_99e/gcuvier .html.

2. Fabrizio Sergio et al., "Ecologically Justified Charisma: Preservation of Top Predators Delivers Biodiversity Conservation," *Journal of Applied Ecology* 43, no. 6 (December 2006): 1049–55.

3. Ritter, "Fact Sheet: Tiger Sharks."

4. "Marine Biologist Carl Meyer Shares Insights on Recent Tiger Shark Bites," University of Hawaii News, October 19, 2015, https://www.hawaii.edu/news/2015/10/19 /marine-biologist-carl-meyer-shares-insights-on-recent -tiger-shark-bites.

5. "Neurotoxins in Shark Fins: A Human Health Concern," ScienceDaily, February 23, 2012, https://www.science daily.com/releases/2012/02/120223182516.htm.

Chapter 6: The Shark Attack Files

1. "Yearly Worldwide Shark Attack Summary," International Shark Attack Files, Florida Museum of Natural History, February 15, 2019, https://www.floridamuseum.ufl.edu /shark-attacks/yearly-worldwide-summary.

2. "Lifetime Odds of Death for Selected Causes, United States, 2017," Injury Facts, accessed January 30, 2019, https://injuryfacts.nsc.org/all-injuries/preventable-death-overview/odds-of-dying.

3. Jaymi Heimbuch, "11 Animals More Likely to Kill You Than Sharks," Mother Nature Network, March 6, 2018, https://www.mnn.com/earth-matters/animals/stories/11-animals-more-likely-to-kill-you-than-sharks.

4. "Species Implicated in Attacks: 1580–Present," Florida Museum, n.d., https://www.flmnh.ufl.edu/fish/isaf/contributing-factors/species-implicated-attacks.

5. Ed Killer, "Shark, Surfer Collide at Fort Pierce Inlet," TCPalm, February 26, 2018, https://www.tcpalm.com/story/sports/2018/02/26/surfer-wipes-out-lands-inches-shark/373560002.

6. H. David Baldridge, *Shark Attack* (New York: Berkley Medallion Books, 1975).

7. Baldridge, *Shark Attack*.

8. Richard Ellis, *Shark Attack: Maneaters and Men* (New York: Open Road Media, 2013).

9. Bradley M. Wetherbee, Christopher G. Lowe, and Gerald L. Crow, "A Review of Shark Control in Hawaii with Recommendations for Future Research," *Pacific Science*, 48, no. 2 (April 1994): 95–115.

10. "Unintentional Drowning: Get the Facts," Centers for Disease Control and Prevention, accessed February 6, 2019, https://www.cdc.gov/homeandrecreationalsafety/water-safety/waterinjuries-factsheet.html.

Chapter 8: Bearing Witness

1. "Pacific Ocean" in *Encyclopaedia Britannica* (Chicago: Encyclopaedia Britannica, Inc., 2006).

2. Trevor Sutton and Avery Siciliano, "Making Reform a Priority for Taiwan's Fishing Fleet," Center for American Progress, January 8, 2018, https://www.americanprogress .org/issues/security/reports/2018/01/08/444622/making -reform-priority-taiwans-fishing-fleet.

3. Sutton and Siciliano, "Making Reform a Priority for Taiwan's Fishing Fleet."

4. James X. Morris, "The Dirty Secret of Taiwan's Fishing Industry," *The Diplomat*, May 18, 2018, https://thedip lomat.com/2018/05/the-dirty-secret-of-taiwans-fishing -industry.

5. "Netting Billions: A Global Valuation of Tuna," Pew Charitable Trusts, May 2, 2016, https://www.pewtrusts.org/en /research-and-analysis/reports/2016/05/netting-billions-a -global-valuation-of-tuna.

6. "News Update: U.S. per Capita Seafood Consumption Up in 2017," Seafood Nutrition Partnership, December 13, 2018, https://www.seafoodnutrition.org/seafood-101/news /news-update-u-s-per-capita-seafood-consumption-up -in-2017.

7. Amanda Hamilton et al., *Market and Industry Dynamics in the Global Tuna Supply Chain* (Solomon Islands: Pacific Islands Forum Fisheries Agency, 2011), https://www .researchgate.net/publication/259442804_Market_and

_Industry_Dynamics_in_the_Global_Tuna_Supply_
Chain.

Chapter 9: Human Trafficking at Sea

1. "Cambodia GDP per Capita, 1993–2017, Yearly, USD, National Institute of Statistics," CEIC Data, accessed March 11, 2019, https://www.ceicdata.com/en/indicator /cambodia/gdp-per-capita.
2. Rina Chandran, "Bank Accounts for Fishermen in Thailand Can Help End Abuses, Officials Say," Reuters, March 7, 2018, https://www.reuters.com/article/us-thailand-fishing -slavery/bank-accounts-or-fishermen-in-thailand-can -help-end-abuses-officials-say-idUSKCN1GK0RC.
3. US State Department, *Trafficking in Persons Report: June 2018*, https://www.state.gov/documents/organization /282798.pdf.

Chapter 10: Water, Water Everywhere and Not a Tiger in Sight

1. Ed Yong, "The Fish That Makes Other Fish Smarter," *The Atlantic*, March 7, 2018, https://www.theatlantic.com /science/archive/2018/03/the-fish-that-makes-other-fish -smarter/554924.
2. "Coral Reefs 101," Coral Reef Alliance, n.d., https://coral .org/coral-reefs-101.

3. Chris Perry et al., "Linking Reef Ecology to Island Building: Parrotfish Identified as Major Producers of Island-Building Sediment in the Maldives," *Geology* 43, no. 6 (May 26, 2015), doi: 10.1130/G36623.1.

4. Perry et al., "Linking Reef Ecology to Island Building."

5. Kemal Atlay, "Weakened Coral Reefs Are Facing the Threat of Algae Colonisation," *Science*, April 27, 2016, https://www.sbs.com.au/topics/science/nature/article/2016/04/27/weakened-coral-reefs-are-facing-threat-algae-colonisation.

6. Jonathan L. W. Ruppert et al., "Caught in the Middle: Combined Impacts of Shark Removal and Coral Loss on the Fish Communities of Coral Reefs," *PLOS ONE* 8, no. 9 (September 18, 2013), https://journals.plos.org/plosone/article?id=10.1371/journal.pone.0074648.

7. H. Cesar, P. Vanbeukerling, and S. Prince, "An Economic Valuation of Hawaii's Coral Reefs," report prepared for the Hawaii Coral Reef Initiative Research Program, Honolulu, Hawaii, 2002.

8. "Wolf Reintroduction Changes Ecosystem," My Yellowstone Park, June 21, 2011, https://www.yellowstonepark.com/things-to-do/wolf-reintroduction-changes-ecosystem.

9. "Wolf Reintroduction Changes Ecosystem," My Yellowstone Park.

10. "Sharks: Meet the Seagrass Protectors," National Science Foundation, July 26, 2017, https://www.nsf.gov/discoveries/disc_summ.jsp?cntn_id=242613; Pamela Reynolds, "Seagrass and Seagrass Beds," Smithsonian, April 2018,

https://ocean.si.edu/ocean-life/plants-algae/seagrass-and -seagrass-beds; Egbuche Christian Toochi, "Carbon Sequestration: How Much Can Forestry Sequester CO_2?," *Forestry Research and Engineering: International Journal* 2, no. 3 (May 21, 2018), doi: 10.15406/freij.2018.02 .00040.

11. James W. Fourqurean et al., "Seagrass Ecosystems as a Globally Significant Carbon Stock," *Nature Geoscience* 5 (2012): 505–9, https://www.nature.com/articles/ngeo1477.

12. Shayla Brooks, "How Seagrass Can Stop Climate Change," Green Future, May 5, 2017, https://greenfuture.io/nature /seagrass-facts.

13. Joleah B. Lamb et al., "Seagrass Ecosystems Reduce Exposure to Bacterial Pathogens of Humans, Fishes, and Invertebrates," *Science* 355, no. 6326 (February 17, 2017): 731–33.

14. Lamb et al., "Seagrass Ecosystems."

15. Alejandro Frid, Gregory G. Baker, and Lawrence M. Dill, "Do Shark Declines Create Fear-Released Systems?," *Oikos* 117, no. 2 (February 2008): 191–201.

16. Frid, Baker, and Dill, "Shark Declines."

17. Wes Pratt citing "Blue Waters of the Bahama: An Eden for Sharks," *National Geographic*, March 2007.

18. "Ecotourism: Dollars and Sense," Shark Savers, n.d., http:// www.sharksavers.org/en/education/the-value-of-sharks /sharks-and-ecotourism.

19. Australian Institute of Marine Science, "19 April 2012 Shark Dive Tourism in Fiji Worth US$42.2 Million a Year," Aus-

tralian Government, April 19, 2012, https://www.aims.gov
.au/19-april-2012.

20. G. M. S. Vianna et al., *The Socio-economic Value of the Shark-Diving Industry in Fiji*, Australian Institute of Marine Science (Perth: University of Western Australia, 2011), 20, https://www.pewtrusts.org/~/media/legacy/up loadedfiles/peg/publications/report/shrkfijieconomicre portfinalpdf.pdf.

21. Shannon Service, "Palau's Plans to Ban Commercial Fishing Could Set Precedent for Tuna Industry," *Guardian*, March 26, 2014, https://www.theguardian.com/sustainable -business/palau-sharks-ban-commercial-fishing-tuna -industry.

22. "Palau Bans Commercial Fishing," *Samoa News*, accessed March 2018, https://www.theguardian.com/sustainable-busi ness/palau-sharks-ban-commercial-fishing-tuna-industry.

23. "Palau Bans Commercial Fishing," *Samoa News*.

24. Austin Gallagher and Neil Hammerschlag, "Global Shark Currency," *Current Issues in Tourism* 14, no. 8 (November 2011), doi: 10.1080/13683500.2011.585227.

25. Conservation International, *Economic Values of Coral Reefs, Mangroves, and Seagrasses: A Global Compilation* (Arlington, VA: Center for Applied Biodiversity Science, Conservation International), https://www.icriforum.org/sites/default /files/Economic_values_global%20compilation.pdf.

26. Aldo Leopold, A Sand County Almanac (New York: Oxford Univ. Press, 1968), https://www.aldoleopold.org /teach-learn/green-fire-film/about-green-fire.

27. Leopold, *A Sand County Almanac,* https://www.aldoleo pold.org/wp-content/uploads/2018/04/Why-Hunt_E -Preview-Only_sm.pdf.

Chapter 11: The High Seas

1. Boris Worm et al., "Global Catches, Exploitation Rates, and Rebuilding Options for Sharks," *Marine Policy* 40 (2013): 194–204, http://wormlab.biology.dal.ca/publication/view /worm-etal-2013-global-catches-exploitation-rates-and -rebuilding-options-for-sharks.

2. Food and Agriculture Organization of the United Nations, *The State of World Fisheries and Aquaculture, 2018: Meeting the Sustainable Development Goals* (Rome: FAO, 2018), http://www.fao.org/3/i9540en/I9540EN.pdf.

3. Food and Agriculture Organization of the United Nations, *The State of World Fisheries and Aquaculture, 2018.*

4. "Sector Trends Analysis: Fish Trends in China," Agriculture and Agri-Food Canada, October 2017, http:// www.agr.gc.ca/resources/prod/Internet-Internet/MISB -DGSIM/ATS-SEA/PDF/sta_fish_trends_china_ats_ten dances_poisson_chine_2017a-eng.pdf.

5. Food and Agriculture Organization of the United Nations, *The State of World Fisheries and Aquaculture, 2018.*

6. Chris Mooney and Brady Dennis, "New Maps Show the Utterly Massive Imprint of Fishing on the World's Oceans," *The Washington Post,* February 22, 2018, https://

www.washingtonpost.com/news/energy-environment/wp
/2018/02/22/new-maps-show-the-utterly-massive-im
print-of-fishing-on-the-worlds-oceans/?utm_term=.cla
2f20194f5.

7. R. Blomeyer et al., "The Role of China in World Fisheries," European Parliament, Directorate-General for Internal Policies, 2012.

8. Sarah Gibbens, "High Seas Fishing Isn't Just Destruc-
tive—It's Unprofitable," *National Geographic,* June 6,
2018, https://news.nationalgeographic.com/2018/06/high
-seas-fishing-subsidies-oceans-science.

9. "Estimated Average Annual Catch by Region, in Metric
Tons, 2000–2011," Africa Center for Strategic Studies,
https://africacenter.org/wp-content/uploads/2017/06
/Where-Chinese-Vessels-Fish-cc.png.

10. Prashanth Parameswaran, "Indonesia's War on Illegal
Fishing Nets New China Vessel," *The Diplomat,* December 6, 2017, https://thediplomat.com/2017/12/indonesias
-war-on-illegal-fishing-nets-new-china-vessel.

11. Christina Vallianos et al., "Sharks in Crisis," Wild Aid
(2018), https://wildaid.org/wp-content/uploads/2018/02
/WildAid-Sharks-in-Crisis-2018.pdf.

12. Felix Dent and Shelley Clarke, "State of the Global Market
for Shark Products," FAO Fisheries and Aquaculture Technical Paper No. 590 (Rome: FAO UN, 2015), 5, www.fao
.org/3/a-i4795e.pdf.

13. Dent and Clarke, "State of the Global Market," 7.

14. Dent and Clarke, "State of the Global Market," 97.

15. Dent and Clarke, "State of the Global Market," 125.

16. Dent and Clarke, "State of the Global Market," 1.

17. "Pelagic Sharks: Predators of the High Seas," World Wide Fund for Nature, https://sharks.panda.org/conservation-focus/sharks-and-rays.

18. Michael E. Byrne et al., "Satellite Telemetry Reveals Higher Fishing Mortality Rates Than Previously Estimated, Suggesting Overfishing of an Apex Marine Predator," *Proceedings of the Royal Society B* 284, no. 1860, https://doi.org/10.1098/rspb.2017.0658.

19. "NSU's GHRI Study Shows Ocean's Fastest Shark Is Being Threatened by Over Fishing," NSU Newsroom, Nova Southeastern University, https://nsunews.nova.edu/the-oceans-fastest-shark-is-being-threatened-by-over-fishing.

20. "The Ocean's Fastest Shark Is Being Threatened by Overfishing," ScienceDaily, August 7, 2017, https://www.sciencedaily.com/releases/2017/08/170807112842.htm.

21. World Wide Fund for Nature, "21st Special Meeting of the International Commission for the Conservation of Atlantic Tunas (ICCAT)," Position Paper 2018, November 12–19, 2018, http://d2ouvy59p0dg6k.cloudfront.net/downloads/wwf_pp_iccat_2018.pdf.

22. Enric Sala et al, "The Economics of Fishing the High Seas," *Science Advances* 4, no. 6 (June 6, 2018), http://advances.sciencemag.org/content/4/6/eaat2504.full.

23. Food and Agriculture Organization of the United Nations, *The State of World Fisheries and Aquaculture, 2018.*

Chapter 12: Shark Warriors

1. Amelia Meyer, "Ragged Tooth Shark," SharksInfo, 2013, http://www.sharksinfo.com/ragged-tooth-shark.html.

2. Beautiful News, "From Sharkbait to Shark Saviour," East Coast Radio, interview, July 20, 2018, https://www.ecr .co.za/shows/stacey-norman/sharkbait-shark-saviour.

3. Adam Welz, "Shark Mystery: Where Have South Africa's Great Whites Gone?," *Yale Environment 360*, August 6, 2018, https://e360.yale.edu/features/shark-mystery-where -have-south-africas-great-whites-gone.

4. "The Republic of South Africa," UN Food and Agriculture Organization, November 2005, http://www.fao.org/fi/old site/FCP/en/ZAF/profile.htm.

5. Tamzyn Zweig, "Educating Anglers to Save Sharks," Save Our Seas, September 25, 2012, https://saveourseas.com /update/educating-anglers-to-save-sharks.

6. "Scientists Predict Major Shifts in Pacific Ecosystems by 2100," Stanford News, September 24, 2012, https://news .stanford.edu/news/2012/september/pacific-ecosystem -shift-092412.html.

Chapter 13: Shark Alley

1. Amelia Meyer, "Sharks—Teeth," 2013, https://www.sharks info.com/teeth.html.
2. Ben Harding, "Marine Scientists Scratch Heads over Sardines," IOL, August 30, 2004, https://www.iol.co.za /news/south-africa/marine-scientists-scratch-heads-over -sardines-220684.
3. D. J. Wuebbles et al., eds., *Climate Science Special Report: Fourth National Climate Assessment*, vol. 1 (Washington, DC: U.S. Global Change Research Program, 2017), https:// science2017.globalchange.gov/downloads/CSSR2017_Full Report.pdf. This report was written by personnel from thirteen federal departments and agencies, including NOAA, NASA, and the DOE.
4. Wuebbles et al., *Climate Science Special Report*, 25.
5. Wuebbles et al., *Climate Science Special Report*, 14.

Chapter 14: Save the Shark

1. Arthur Koestler, *The Ghost in the Machine* (Last Century Media, 1982), 305.
2. "Fishermen Caught Trying to Smuggle Shark Fins from Hawaii Plead Guilty," Hawaii News Now, December 14, 2018, http://www.hawaiinewsnow.com/2018/12/14/indone sian-fishermen-caught-trying-smuggle-shark-fins-hawaii -get-off-with-fine.

3. "Webster Introduces Commonsense Solution for Shark Trade and Conservation," Daniel Webster Press Release, March 15, 2018, https://webster.house.gov/2018/3/webster -introduces-commonsense-solution-for-shark-trade-and -conservation.

4. "Shark Fin Trade: Why It Should Be Banned in the United States," OCEANA, n.d., https://usa.oceana.org /sites/default/files/4046/shark_fin_ban_announcement _brochure_final_low-res.pdf.

HARPER LUXE

THE NEW LUXURY IN READING

We hope you enjoyed reading
our new, comfortable print size and found it
an experience you would like to repeat.

Well — you're in luck!

HarperLuxe offers the finest in fiction and
nonfiction books in this same larger print size and
paperback format. Light and easy to read, HarperLuxe
paperbacks are for book lovers who want to see
what they are reading without the strain.

For a full listing of titles and
new releases to come, please visit our website:

www.HarperLuxe.com